D0203462

# *Race to the Moon*

**Also by William B. Breuer**

An American Saga

Bloody Clash at Sadzot

Captain Cool

They Jumped at Midnight

Drop Zone Sicily

Agony at Anzio

Hitler's Fortress Cherbourg

Death of a Nazi Army

Operation Torch

Storming Hitler's Rhine

Retaking the Philippines

Devil Boats

Operation Dragoon

The Secret War with Germany

Sea Wolf

Hitler's Undercover War

Geronimo!

Hoodwinking Hitler

# RACE TO THE MOON

## *America's Duel with the Soviets*

WILLIAM B. BREUER

Westport, Connecticut
London

**Library of Congress Cataloging-in-Publication Data**

Breuer, William B.
Race to the moon : America's duel with the Soviets /
William B. Breuer.
p. cm.
Includes bibliographical references and index.
ISBN 0–275–94481–6 (alk. paper)
1. Space flight to the moon—History. 2. Project Apollo (U.S.)
3. Rocketry—Research—Germany—History. I. Title.
TL799.M6B725 1993
629.45'4'0973—dc20 92–31849

British Library Cataloguing in Publication Data is available.



Library of Congress Catalog Card Number: 92–31849
ISBN: 0–275–94481–6

First published in 1993

Praeger Publishers, 88 Post Road West, Westport, CT 06881
An imprint of Greenwood Publishing Group, Inc.

Printed in the United States of America

The paper used in this book complies with the Permanent Paper Standard issued by the National Information Standards Organization (Z39.48–1984).

10 9 8 7 6 5 4 3 2 1

Dedicated to

MONSIGNOR FRANCIS L. SAMPSON

Major General (Ret.), U.S. Army

A paratrooper padre who, in
two wars, leaped behind enemy lines
and into the midst of combat
with no weapon other than
the sword of the spirit.
His battlefield exploits and
escapes from captivity are legendary.
For 30 illustrious years he
served God and country.

What we attained when Neil Armstrong stepped down on the moon was a completely new step in the evolution of man. It caused a new element to sweep across the face of this good earth and to invade the thoughts of all men. For the first time, life left its planetary cradle and the ultimate destiny of man is no longer confined to these familiar continents that we have known so long.

—Wernher von Braun

# Contents

# Acknowledgments

In collecting research materials for this book, I tracked down and interviewed or contacted 112 persons who were involved in the story. Many chose to remain anonymous, but their contributions were of great value. In some instances, they were the widows of participants.

For their unflinching help in tapping their sources to find many of these individuals, I express my appreciation to Lieutenant General William P. Yarborough, Lieutenant General Edward M. Flanagan, Jr., Lieutenant Colonel Haynes W. Dugan, Colonel William A. Castille, Lieutenant Colonel William L. Howard, and Arthur A. Klekner, all retired U.S. Army officers.

Also helpful in locating key participants were John S. Welch, Jr., chairman of the board of General Electric Company in Fairfield, Connecticut, whose corporation played a highly significant role in Operation Overcast. Thanks also to Carla M. Fischer, public relations representative at General Electric Company, for providing valuable information; and to Ruth Shoemaker of the General Electric Hall of History in Schenectady, New York.

Other business executives whose suggestions and guidance were beneficial included Kenneth A. Roe, chairman of the board of Burns and Roe Enterprises in Oradell, New Jersey; and William Carlson, former aerospace director at Burns and Roe.

Appreciation is expressed to U.S. Senator Edward M. ''Ted'' Kennedy for steering the author toward certain materials with regard to President John F. Kennedy's challenge to put a man on the moon. Likewise to Paul B. ''Red'' Fay, Jr., of San Francisco and Vice Admiral John D. Bulkeley (Ret.) of Silver Spring, Maryland, both of whom were longtime close friends of President Kennedy and provided valuable insights into his thinking. (Bulkeley retired from the Navy in 1989, after 56 years of service, as the United States most highly decorated combat man.)

In Washington, D.C., Lee D. Saegesser, archivist at the National Aeronautics and Space Administration (NASA), provided the author with significant research materials and contacts. And the staff of the National Archives was most helpful.

Also in Washington, I thank Brigadier General Charles W. McClain and Colonel Michael V. Sullivan, in the Office of Public Affairs at the Pentagon, and Inspector Thomas F. Jones, in the Office of Public Affairs at the Federal Bureau of Investigation.

In Huntsville, Alabama (Rocket City, USA), Walter Wiesman, the youngest of the Peenemünde team that came to the United States in 1945, helped the author in a variety of ways, including providing a chronology (1927–1980) of the German rocket team that he had co-authored with another Huntsville resident, Ruth G. von Saurma.

Most helpful also was Michael E. Baker, chief historian at the United States Army Missile Command, Redstone Arsenal, in Huntsville, who provided much background information and other research materials. Helen Berrisford, research assistant at the Huntsville/Madison County Chamber of Commerce, and John B. Taylor, director of public affairs at the George C. Marshall Space Center in Huntsville, contributed valuable assistance.

Staff members of numerous public and private libraries were of great assistance to me in finding pertinent research materials. They included: Charles Steinhice at the Chattanooga Bicentennial Library; Benedict K. Zobrist, director, and George H. Curtis, archives assistant, at the Harry S Truman Library in Independence, Missouri; Ronald E. Whealan, reference librarian, and June Payne, archives assistant, at the John Fitzgerald Kennedy Library in Boston; and Adeline Collins, librarian, and Dion Sprague and Ann Draper at the Cleveland (Tennessee) State Community College Library.

In Great Britain, Reginald V. Jones, one of that nation's most brilliant scientists while still in his 20s during World War II, was most helpful to me with his recollections concerning the efforts of British intelligence to learn details of the German long-range missile program. Thanks also go to Major G. G. Norton, curator of the Airborne Museum in Aldershot, England, and to Hilary Roberts and her associates at the Imperial War Museum in London.

Finally, appreciation is expressed to my wife, Vivien, who translated the many German-language documents that played an important role in the creation of this book.

*William B. Breuer*

*Race to the Moon*

# 1

# "This Great New American Enterprise"

Speeding down Washington's wide Independence Avenue behind a covey of motorcycle policemen, the hunched, tousle-haired figure in the rear of the black Cadillac limousine ignored not only the noontime crowds but also the White House aides who rode with him. Forty-three-year-old President John Fitzgerald Kennedy had little time to think of anything but the looseleaf notebook on his lap. It was May 25, 1961.

Snapped into the notebook were the pages of a speech, which, Kennedy knew, would prove to be a blockbuster. The president had been working intermittently on the text for three weeks, and he was still hacking out paragraphs, altering words and penciling in last-minute thoughts.

The presidential entourage was bound for Capitol Hill where Kennedy would perform the unique ritual of presenting a second State of the Union message to a joint session of Congress, less than four months after he had given the traditional first one. This time, Kennedy would present to Congress a score of proposals, ranging from expanding domestic social programs to beefing up the armed forces. His most dramatic—and costly—proposal would be a request for perhaps $40 billion over the next ten years to put an American on the moon and return him safely to earth.

Sending an American to land on the lunar surface was an urgent crusade for idealistic, energetic Jack Kennedy. Like most Americans, the president was chagrined and frustrated over a series of Russian space spectaculars. Only recently, he had reached a decision: the United States would not give up on the manned "space race," even though the Soviets had a commanding lead that many American scientists were convinced could not be overcome.

America's global prestige, Kennedy concluded, could not endure a second-place finish in man's quest for the moon. He was convinced that the Dwight Eisenhower administration and Americans as a whole had not and did not fully grasp the worldwide political and psychological impact of the space race with the Soviets. Since the conclusion of World War II in 1945, the United States and the Soviet Union had been competing vigorously to convince the world, especially the new and undecided Third World nations, which way to turn, which was the wave of the future—a free enterprise system like American capitalism or a regimented structure like Soviet communism.

Kennedy, a Harvard University graduate and World War II PT boat skipper in the Pacific, always had a highly competitive nature. With his older brother, Joseph P. Kennedy, Jr.,[1] he won the intercollegiate sailboat championship, and he excelled on the Harvard swimming team. Although he was too light to have much chance of making the Harvard varsity football team, he gave it a try, and during scrimmage in his sophomore year, he suffered a spinal injury that would later threaten both his political career and his life.

"No matter what kind of competition Jack was engaged in—war, sports, or politics-he bent all of his energies and skills to winning," recalled Paul B. "Red" Fay, Jr., who had been one of Kennedy's closest friends since they served together in PT boats in World War II. "Jack felt that every generation of every great nation had their challenge or challenges," Fay added. "If they failed to respond to the opportunities of their time, their claim to greatness would diminish. Going into space was the challenge of his time, so reaching the moon became his frontier."[2]

During his heated 1960 presidential campaign against Republican Richard M. Nixon, Kennedy had hit hard and repeatedly at the "space gap" he said that the Eisenhower-Nixon administration had "allowed." Kennedy told audiences across the land that the space gap symbolized the nation's lack of initiative, ingenuity, and vitality under Republican rule. Privately, Kennedy made no bones of the fact that he considered 70-year-old Eisenhower, who was completing his second term, to be both "tired" and "shallow" and lacking in the creation of innovative goals.[3]

In campaign speeches, Jack Kennedy dramatized the Soviet frontrunner position in the space race by declaring: "The first living creatures to orbit the earth in space were dogs named Strelka and Belka, not Rover or Fido—or even Checkers [Nixon's celebrated canine]."

Time and again, the Democratic contender from Massachusetts reminded audiences that Vice President Nixon had told Nikita Khrushchev, the bellicose Soviet dictator, in a widely publicized, face-to-face confrontation: "You may be ahead of us in rocket thrust, but we are ahead of you in color television." Then Kennedy would deliver his punch line: "I will take my television in black and white. I want to be ahead in rocket thrust!"[4]

It was potent stuff. Enthusiastic crowds loved it and roared their approval. Nixon fumed.

On November 8, 1960, Kennedy defeated Nixon by a razor-thin margin—34,227,096 votes to 34,107,646. While the tally was hardly a mandate from the American people, Kennedy eagerly set into motion plans for implementing his New Frontier ideas. One of his first acts after being sworn into office, in January

1961, was to establish a blue-ribbon task force headed by his scientific advisor, Jerome B. Weisner. Much to Kennedy's dismay, Wiesner and his panel threw cold water on the new president's space goal, concluding that the United States could not win the race with the Soviet Union to put a man on the moon. Russia was simply too far ahead in rocket thrust and other technological know-how.

There was no shortage of space-flight scoffers in the scientific community, at home and abroad. One of these was Vannevar Bush, chairman of the Board of Governors of the prestigious Massachusetts Institute of Technology, who had headed the Office of Scientific Research and Development in World War II. Testifying before the House Committee on Science and Technology in Washington on April 6, 1961, the 70-year-old Bush had sharp criticism. "There is too much hullabaloo over the propaganda aspects of the space program," the noted scientist declared. "Soviet space achievements have merely hurt our pride. . . . Putting a man on the moon is a stunt. Man can do no more than an instrument, in fact can do less. The days when men will be in space for long periods and for varied purposes are so far off that we need not hurry." Bush said that some of the futuristic proposals for space projects were "simply unadulterated absurdity."[5]

Across the Atlantic, the Astronomer Royal of England, Richard van de Riet Wooley, publicly declared that the idea of space travel was "utter bilge." When a Soviet Army major named Yuri Alekseyevitch Gagarin was launched into space on April 12, 1961, and orbited the earth for four and a half days, members of the British Interplanetary Society gleefully reported that Gagarin had penetrated "utter bilge."

Gagarin's stunning feat—the first human in space—electrified the world and provided the Soviet Union with an enormous propaganda tool with which to trumpet the superiority of the Communist system over America's free enterprise system. In a jab at the United States, Nikita Khrushchev loudly proclaimed Major Gagarin to be "the new Christopher Columbus."

President Kennedy was among the millions of Americans who were crestfallen. "It'll take some time to catch up to the Soviets," a somber Kennedy told the media. "We are, I hope, going to go in other areas where we can be first, and which will bring perhaps long-range benefits to mankind. But we are behind."[6]

Kennedy fired off a message of congratulations to Premier Khrushchev, then began a frantic series of conferences with Jerome Wiesner and other space advisors. Everyone agreed that something had to be done—fast. American know-how was being ridiculed around the world for not being able to launch a man into space. Jack Kennedy was more determined than ever to put an American on the moon, ahead of the Russians.

Less than three weeks after Yuri Gagarin had returned to earth, a glaring *New York Times* headline screamed: "U.S. HURLS MAN 115 MILES INTO SPACE." Thirty-seven-year-old Commander Alan B. Shepard, Jr., a slim and confident Navy test pilot, had become the first American space explorer. Shepard landed safely 304 miles out at sea, 15 minutes after launching from Cape Canaveral, Florida.

"Boy, what a ride!" Shepard enthused after being lifted from his bobbing capsule into a hovering Marine Corps helicopter.

Shepard's near-perfect flight was not anywhere near as impressive as the Soviets' had been, but Uncle Sam had at least vaulted out of the starting blocks and had gotten into the space race to catch Ivan the Bear.

Now, on May 25, 1961, President Kennedy was standing before a joint session of Congress to deliver his dramatic challenge to Americans. Speaking in his crisp New England accent and jabbing the air with his forefinger for emphasis, Kennedy declared:

> I believe that this nation should commit itself to achieving the goal, before this decade is out, of landing a man on the moon and returning him safely to earth. No single space project in this period will be more impressive to mankind, or more important for the long-range exploration of space; and none will be so difficult or expensive to accomplish. . . .
>
> In a very real sense, it will not be one man going to the moon . . . it will be an entire nation. For all of us must work to put him there. . . .
>
> No one can predict with certainty what the ultimate meaning will be of mastery of space.
>
> I believe we should go to the moon.[7]

Seated at the side of the rostrum, Theodore C. Sorensen, Kennedy's close friend and confidant, thought the president looked strained in his effort to win over the legislators. Sorensen noticed that the president suddenly departed from his prepared text—the only time he had ever done so in addressing Congress—to declare: "Unless we are prepared to do the work and bear the burdens to make it successful, there is no sense in going ahead." Sorensen reflected that Kennedy's voice, while urgent, sounded a little uncertain.[8]

At the conclusion of Kennedy's 47-minute message, which was televised "live" by the CBS, NBC, and ABC networks, there was only routine applause. As they left the crowded House floor after the president departed, members of Congress were split over his proposal to spend $40 billion in the decade ahead in an all-out crusade to beat the Russians to the moon.

What the Republicans and the conservative Southern Democrats derided as Kennedy's "huge spending schemes," the Democrat liberals described as "necessary sacrifices." Fears were expressed that the huge outlay of funds might divert money from social programs such as aid to the elderly. But one fact was certain: President Kennedy had lobbed the ball into Congress' side of the court.

Riding back to the White House, Kennedy remarked to Ted Sorensen that the applause that greeted his proposal to go to the moon was "hardly enthusiastic."[9] Despite the president's gloomy reading of the mood on Capitol Hill, he proved to be a consummate salesman. Congress bought his pitch for "this great American enterprise" and hiked the space budget by 50 percent for that year. The next year the space budget would be more than the combined funding for space exploration in all the years prior to mid–1961.

Landing a man on the moon first was a symbol of the continuing struggle for the minds of men between the United States and the Soviet Union, but no one could be certain that ultimate success would prevail. President Kennedy had committed the United States to a mighty endeavor with all possible speed and colossal expenditure of money and human resources.

Yet the chain of events that would permit the United States to reach ever higher into the unexplored heavens began 22 years earlier, not in the United States, but in Oslo, Norway.

# 2

# The Oslo Mystery Package

Early in the morning of October 17, 1939, an aide walked into the office of Vice Admiral Hector Boyes, the Royal Navy attaché at the British embassy in Oslo, and handed him a sealed envelope that had just arrived in the regular Norwegian mail delivery system. These were tense days in the embassy. At 4:30 A.M. on September 1, just over six weeks earlier, 59 German divisions, paced by swarms of shrieking Stuka dive-bombers, plunged over the border into Poland, converged on Warsaw from three sides, and conquered the militarily weak country in only 27 days. The speed, power, and finesse of Adolf Hitler's legions, the most powerful war juggernaut that history had known, created a new word in the languages of many nations: *blitzkrieg* (lightning war).

When the German fuehrer curtly rejected an ultimatum by British Prime Minister Neville Chamberlain to withdraw, England and France, on September 3, declared war on Germany. Great Britain, woefully unprepared for a major conflict, and France could only sit on the sidelines and watch the far-outnumbered and ill-equipped Polish Army be smashed.[1]

Now, in Oslo, three weeks after the Nazi swastika began flying over Warsaw, Admiral Boyes, a slim, energetic officer, ripped open the envelope he had just been handed and began reading the anonymous letter enclosed. Written in longhand, the source said that, if the British wanted highly important information on German technical and weapons developments, they should indicate an affirmative answer by altering the beginning of the regular BBC (British Broadcasting Company) German Service broadcast to insert the words: "*Hullo, hier ist London*" (Hello, here is London).[2]

Intrigued and mystified by the curious letter from an unknown yet seemingly authoritative source, Boyes rushed the missive to his government in England.

London, in late autumn 1939, was gripped by an atmosphere of foreboding. There had been plenty of war news—all of it bad. On the night of October 13–

14, a daring German U-boat (submarine) skipper had slipped his craft into the narrow passage of Scapa Flow, in northern Scotland, and sent two torpedoes into the mighty British battleship *Royal Oak*. The huge vessel sank to the bottom, taking 806 officers and crewmen to watery graves.

It was in this climate of gloom that Stewart Menzies, the *C* (Chief) of MI–6, Great Britain's secret intelligence service, and his top officials gathered at his headquarters on Broadway, a quiet side street near Parliament Square and Westminster Abbey. Behind Menzies' desk was a portrait of his patron, the late King Edward VII, dressed in hunting tweeds, a shotgun in one hand and a hunting dog beside him.

The 49-year-old, distinguished-looking Menzies was born into a wealthy British ruling-class family. His father's riches came originally from whiskey and gin distilling; his mother's family owned the Wilson Steamship Company. Despite his lofty perch in Britain's nobility, Stewart Menzies had spent most of his adult life in the often sinister cloak-and-dagger machinations of His Majesty's secret service.

C and the others were intensely discussing the significance of the anonymous letter that Boyes had received in Oslo two days earlier. Was it the handiwork of a crackpot? Several felt that it might be some sort of psychological warfare gambit perpetrated by the *Abwehr*, the Third Reich's secret service, a clandestine apparatus with some 3,000 agents on seven continents around the world.

Menzies and the others knew that the Abwehr was headed by one of history's most cunning masterspies, Admiral Wilhelm Canaris, a small, nervous man with an intense disposition who had been engaged in undercover operations for Germany since World War I. Only Menzies knew, however, that Canaris was playing a double-game: while being one of the most important cogs in Adolf Hitler's war machine, the 53-year-old Abwehr chief also was a ringleader in the *Schwarze Kapelle* (Black Orchestra), a small, tightly knit group of German admirals and generals, government officials, and civic leaders who for several years had been conspiring to overthrow (or assassinate) the fuehrer and destroy the Nazi regime. Hitler, the German conspirators were convinced, was taking the Fatherland hell-bent down the road to destruction.

For many months, Canaris had been in clandestine contact with Stewart Menzies and had been feeding him a stream of reports on the military and political situation in Germany. Consequently, Menzies may have been convinced that the Abwehr boss was indeed the instigator of the Oslo mystery letter, but he had to remain silent.[3]

Finally, the MI–6 leaders, at Menzies' suggestion, decided to follow up on the Oslo letter and authorized changing the BBC broadcast preamble to ''*Hullo, hier ist London*.''

On November 4, 1939, a guard at the British embassy in Oslo was making his rounds in a heavy snowstorm when he spotted a parcel on a stone ledge. Wrapped in kraft paper and bound by sturdy string, the container was roughly three inches thick, 12 inches wide, and 15 inches long. It was half covered by snow; had it not been found at the time, the package might have been buried in the snow and not detected for weeks.

The parcel was addressed to Admiral Hector Boyes, and the guard rushed it to the naval attaché. When Boyes opened the package—very gingerly, for it might have contained a bomb—he found eight pages of typewritten text detailing German technical innovations being developed and a number of sketches of what appeared to be revolutionary new weaponry, including huge rockets.

Boyes rushed the package to London.

Thirty-six hours after the British embassy guard discovered the mystery package, copies of its contents were distributed to several English scientists and intelligence experts for their analyses and conclusions. Although the contents aroused much excitement, most scientists were skeptical. One physicist, Julius Mader, was convinced that the author of the Oslo document was Hans Heinrich Kummerow, who was a key leader in *Die Rote Kapelle* (the Red Orchestra), the Communist spy ring in Germany.[4]

Whoever may have been the author, how had one man obtained access to such top-secret German materials, then presented them to the British in such concise fashion? A few of the scientists considered the documents to be fakes, a devious scheme by the Germans to cause the British to waste valuable time pursuing technological developments that did not exist.

One scientist, Reginald V. Jones, eagerly examined the Oslo parcel's contents in his office in London's MI–6 headquarters. Jones, 28 years of age, had been educated at Oxford and was on the staff of the secret intelligence service. Despite his relative youth, Jones had a reputation as one of England's most brilliant scientific brains.

Typical of those in his calling, Jones approached his examination with skepticism. But he soon concluded that the mystery writer who assembled the detailed report had an extensive technical and scientific background and, obviously, had direct access to weapons developments in the Third Reich.

Jones was startled by the extent of the disclosures, all of which had plausible scientific logic. They revealed that the Germans were developing huge, long-range rockets and radio-controlled glider bombs at a remote locale on the Baltic coast of the Third Reich—Peenemünde.

Jones' report on his findings in the Oslo mystery package sent shock waves reverberating through the upper levels of British officialdom. If true—a big if—the German scientists were developing an entire new dimension of warfare.

Despite the haunting specter of German long-range rockets that conceivably could bombard London, official British interest waned in what became known as the Oslo Report. Reginald Jones had discovered that Great Britain, a lion at bay, was facing a more immediate crisis: the Germans had developed a revolutionary radio beam by which Hitler's mighty Luftwaffe could be guided to targets in bad weather and in the blackness of night. So Jones and other British scientists focused on creating technology that would foil the beam, called the *X-Gerät* (X-Apparatus), and the Oslo Report was pigeonholed.[5]

Peenemünde, located on the island of Usedom, just off the German mainland on the Baltic coast, near the mouth of the Oder River, was the top-secret site of the *Heeresversuchstelle* (Army Experimental Station). Peenemünde had been a sleepy fishing village until late 1936, when Hitler began expanding his armed forces and weaponry. An airfield was built and a new north-south road split the northern part of Usedom into two parts.

The eastern side was taken over by the *Heer* (Army) to develop massive long-range rockets, and the Luftwaffe had control of the western side where it was developing the V–1 (a pilotless aircraft known later as a buzzbomb). A staff of scientists, physicists, engineers, technicians, and clerical workers eventually would number some 10,000 persons.

In addition to the large force at Peenemünde, many prestigious scientific institutions in Germany were called on to contribute their expertise, including the *Hermann Göring Institut* in Braunschweig, the *Raketenflugtechnische Forschunginstitut* (Institute of Technical Research on Rockets) in Trauen, and the *Deutsche Institut für Luftfahrforschung* (the German Institute for Air Research) in Berlin.

Peenemünde was one of the most tightly guarded facilities in the world. Abwehr agents and picked members of the SS seemed to be everywhere. There was never any mention of Peenemünde or the experiments there in German newspapers or radio stations, all firmly controlled by Adolf Hitler's propaganda genius, diminutive Joseph Goebbels. Each person working at Peenemünde was sworn to complete secrecy under pain of lengthy imprisonment.

Technical director on the army side of Peenemünde was Germany's most brilliant rocket scientist, Baron Wernher von Braun, who was only 27 years old when the Oslo Report fell into the hands of British intelligence in 1939. Tall, sturdily built, and a handsome bachelor, von Braun had a well-developed sense of humor and a flashing smile that set many feminine hearts aflutter.

Wernher von Braun was born March 23, 1912, in Wirsitz, East Prussia, one of three sons of Baron Magnus von Braun, a wealthy descendent of a Prussian family which had served the Fatherland loyally and with distinction for seven centuries. Magnus was a founder of the German Savings Bank, and he brought his family to Berlin, where they took up residence in a spacious mansion. Later, the elder von Braun became minister of agriculture in the Weimar Republic of President Paul von Hindenberg, Germany's World War I hero.

Early on, Wernher von Braun became obsessed with rocketry, and many who knew him held that the boy was "slightly touched in the head" because he often spoke buoyantly of man one day going to the moon. He devoured eagerly such books as Jules Verne's *From the Earth to the Moon* (written in the mid–1860s) and H. G. Wells' *The First Men on the Moon*.

Wernher von Braun's first spaceward influence may have come from his gentle and soft-spoken mother, Baroness Emmy von Braun, who descended from the von Quistorp family of mixed German, French, and Swedish heritage. Early in life, she mastered six languages. For his confirmation, Wernher did not get a watch and his first pair of long pants like most Lutheran boys; he received a telescope from his mother.

Baroness Emmy taught her son to study the planets and the stars through the telescope, and he became a skilled amateur astronomer which led to his interest in the universe. While not yet in his teens, Wernher had an intense interest in designing some type of vehicle for carrying a man to the moon.

In 1930, at 17 years of age, Wernher began experimenting with small rockets, paying his own bills and using a municipal dump on the outskirts of Berlin. He and his fellow enthusiasts called the site the *Raketenflugplatz* (Rocket Flight Place). Most of his associates were equally young and amateurish, but one of their mentors was Hermann Oberth, an expert in rocket fuels who had inspired many young German rocketeers.

Oberth was born at Hermannstadt, Transylvania, in June 1894. His father, Julius Oberth, was a physician. The son did college work at the University of Klausenburg, at Göttingen and at Heidelberg. In 1923, Oberth became a professor of mathematics and physics, and that same year a book that was to enjoy widespread recognition, *Rocket to Interplanetary Space*, was published. For two years, 1928 to 1930, Oberth experimented with gasoline and liquid-air rocket propellants, helping to finance his work by serving as an advisor to a film company making *Frau im Mond* (Lady in the Moon).

While Wernher von Braun was continuing his rocket experiments, he graduated from the Berlin Technological Institute in early 1932. Later, he received his doctorate in physics from the Freidrich Wilhelm University of Berlin. His thesis dealt with theoretical and experimental aspects of liquid propellant rocket engines.

Beginning in 1930, German army leaders developed an interest in rockets and appointed Captain Walter Dornberger, then 34 years of age, as chief of the German Board of Ordnance for the development of modern rockets. Dornberger, a professional engineer, had obtained his degree from the Berlin Technical University. The Versailles Treaty, imposed on Germany by the victorious Allies at the conclusion of World War I, had limited the amount of artillery in the 100,000-man army but said nothing about barring rockets.

One day in August 1932, a black sedan halted by the Raketenflugplatz where Wernher von Braun and a few others were testing rockets, and three German army officers in civilian clothes stepped out. One of them, Captain Walter Dornberger, offered 21-year-old von Braun an offer the delighted young man could not refuse: a civil service position with the army in charge of the development of liquid rockets. The boy wonder was on his way in the mystical field of rocketry. To von Braun, it was a golden opportunity to begin space exploration, and he would be paid for doing what he loved to do best. At the time, he gave no thought about the possible use of rockets in war.[6]

Von Braun's first assignment at Kummersdorf West, an army proving ground 17 miles south of Berlin, was to build a rocket and test it. Along with a few assistants, the young scientist worked night and day for a month, then was ready to test his new rocket.

On the bitter cold night of December 31, 1932, the rocket was mounted on a test stand. Von Braun, Dornberger, and others took shielding positions behind nearby trees as a technician set afire the gasoline in the rocket engine before he, too, dashed for cover. Moments later, a blinding flash illuminated the black sky and a thunderous roar echoed over the surrounding pine forests. Then an eerie silence. The rocket had exploded on the test stand.

Smoke enveloped the site as the rocket men sadly surveyed the twisted wreckage of the test stand. Von Braun was crestfallen. Walter Dornberger put a comforting hand on the youth's shoulder. "Don't worry, Wernher," he consoled. "We will have many failures, but from each we learn, and one day we will know great success." Prophetic words.[7]

Von Braun, Dornberger, and the other pioneers persevered and soon developed a rocket that was fired upward for a mile and a half. Joy reigned at Kummersdorf. Inexorably, the technical problems were being isolated and solved, and the rockets were becoming larger and more powerful.

Overcoming technical difficulties was but one problem. Another was the heavy hand of German bureaucracy, an impediment to scientific development in most nations. In Berlin, officials in the Bureau of the Budget regarded the rocketeers as "crazy engineers." Kummersdorf's spending was carefully audited, and the rocketeers were not allowed to purchase needed office equipment or tools.

Undaunted, von Braun used his ample ingenuity to circumvent the eagle-eyed auditors in Berlin. When ordering typewriters, the requisition request was filled out for "instruments for recording test data with rotating roller." Pencil sharpeners were described as "appliances for milling wooden dowels up to ten millimeters in diameter." These requests were approved.

Despite his second son's notable achievements at such a tender age, Baron Magnus von Braun was far from enthralled. Wernher's obsession with rockets and space travel seemed to be utter nonsense, a classic way to waste one's life and talents. Magnus always had hoped that his son one day would take up the dignified duties of a "landed Prussian gentleman."

On January 7, 1936, four thousand miles west of Berlin, across the Atlantic Ocean, 36-year-old Gustav Guellich had just returned from a long and tiresome round-trip to New Mexico. He was seated at a small desk in a room at the Hotel Martinique, on New York's West 32nd Street, where he lived. Guellich, a native of Munich, had come to the United States in 1932 and was a metallurgist in the laboratories of the Federal Shipbuilding Company, a division of U.S. Steel, in

Kearny, New Jersey. An emaciated bachelor who suffered periodic attacks of depression, Guellich kept to himself and carried out his shipyard duties quietly and efficiently. His fellow workers at Kearny considered him to be an odd duck. Guellich was also a German spy.

Gaining access to secret or classified materials from his vantage point in the sensitive shipyard laboratory, Guellich sent to the Abwehr in Germany a flood of high-grade data about U.S. top-secret developments and other American scientific and military matters.

When Adolf Hitler seized power and began rearming Germany in the mid–1930s, he concluded that the United States, with its gigantic industrial potential, would be the ''decisive factor'' in any forthcoming world war, so the fuehrer and his supreme spymaster, Admiral Wilhelm Canaris, clandestinely began invading the United States with a massive espionage network.

The United States of the 1930s was a spy's haven. No single federal government agency had been designed to combat and root out subversive forces; security measures in key military, government, industrial, and scientific agencies were virtually nonexistent.

Gustav Guellich had been recruited into the Abwehr's spy apparatus by Ignatz Theodor Griebl, a prominent New York City physician and womanizer who was head of the German spy network in that region.

Now, in the Hotel Martinique, unobtrusive Gustav Guellich was compiling for delivery to Berlin his greatest coup: a four-page report entitled *Experiments with High-Altitude Rockets in the United States*. It contained details of work that had been done by Robert H. Goddard, professor at Clark University in Worcester, Massachusetts. Goddard, Guellich wrote, had made ''a substantial breakthrough in the development of rocket-propelled missiles.''[8]

A few weeks earlier, learning of Goddard's experiments, Guellich took a trek of some 2,500 miles to a barren locale in New Mexico and watched, from concealment, without interference from anyone, as Goddard and a handful of assistants scored a brilliant success by firing a rocket controlled by a gyroscope and by vanes in the exhaust system. It rose to an altitude of 4,800 feet, swung to horizontal flight in response to its steering mechanism, reached a speed of 550 miles per hour, and continued for nearly three miles before plunging to earth.

Born in 1882, Goddard was among the first engineers in the United States to give serious study to the possibility of rockets. In 1919, he wrote a document called *A Method of Reaching Extreme Altitudes*, which presented formulas for computing the power necessary to put a rocket on the moon. America yawned and branded Goddard as a hare-brained crackpot.

Undaunted, Goddard put in years of hard work and sacrifices to elevate rocket engineering from mathematical theory to practical testing. On May 16, 1926, Goddard launched the world's first rocket powered by a liquid propellant motor. The site of the liftoff was on the outskirts of Auburn, a suburb of Worcester.[9]

Ten years later, Gustav Guellich's missile espionage bonanza created a flap in Berlin, and the spy was pressured for more information on American rockets. Unknown to Guellich, German scientists were also conducting rocket experiments, and Goddard's findings would be an enormous boon to the German efforts.[10]

One afternoon a few months after Guellich's four-page missile report reached the Abwehr in Berlin in mid–1936, there was a knock on his door at the Hotel Martinique. It was Karl Eitel, an Abwehr courier on a cross-Atlantic luxury passenger liner. Eitel brought exciting news—belated recognition. Guellich had been elevated one notch up the Abwehr totem pole, from Number F.2307 (a subagent) to Agent 2336 (a full-fledged spy).

# 3

# "Today the Space Ship Was Born!"

Just past 8:00 A.M. on May 7, 1939, Adolf Hitler was awakened in his apartment in the magnificent new *Reichkanzlie* (Chancellery) in Berlin. Rain was beating down and a cold, biting wind was whipping through the capital city.

The 50-year-old fuehrer of the Third Reich was in a foul mood. Much to his annoyance, he had promised General Walther von Brauchitsch, commander of the German army, that he would watch a rocket-firing demonstration at the army testing grounds at Kummersdorf West, 17 miles south of Berlin.[1] Hitler's grumpy disposition was compounded by the fact that he had had only two hours sleep. Essentially a "night person," the fuehrer was accustomed to holding long and tedious discussions with his military and government leaders from darkness to dawn. These discussions were largely Hitler monologues with the others involved able to inject little more than monosyllabic comments. Then the leader of the Third Reich would sleep most of the day. After a hot bath, some medicine, and injections by his personal physician, Theodor Morell, the fuehrer was ready for another night-long round of discussions, maybe with time out to watch a movie.[2]

By German law, Adolf Hitler's full title was a jawbreaker: *der Fuehrer und Oberste Befehlshaber der Wehrmacht des Grossdeutschen Reichs*. Physically, he was not an imposing figure. Of medium height and build, his straight, jet-black hair was draped over his forehead, and his most dominant facial feature was a brush mustache which he had worn during his entire manhood.

Those who came into close contact with the fuehrer agreed that he had a hypnotic cast to his eyes which often influenced others to want to do his bidding. His personality was a classic enigma. He could be cruel and utterly ruthless, going into a towering rage when events or individuals displeased him. On the other hand, he also had a warm and winning demeanor with those he looked upon with favor.

Hitler was a spellbinding orator who turned massive audiences into fits of frenzy, a trait that helped to propel him into supreme power in a nation of some 90 million people despite having had less than a high school education.

The great day in Austrian-born Adolf Hitler's life was January 30, 1933, when the aged and senile President Paul von Hindenberg appointed him chancellor. Almost immediately, Hitler began cleaning out top officials in the popularly elected Weimar Republic and replacing them with his own loyalists. One of those who ''resigned'' was the minister of agriculture, Baron Magnus von Braun, who retired to the family estate in Selisia.

Now, six years after Hitler had reached the pinnacle of power, he stepped from his limousine at Kummersdorf West and was greeted by a clicking of heels and salutes from assembled brass and rocket scientists.

Walter Dornberger (now a colonel), who had recruited Wernher von Braun, showed diagrams and drawings and gave the fuehrer a briefing on the background of German rocket development. Dornberger, who had become something of a big brother to von Braun, was a rare combination of scientific brains and practical administrator, who, in many ways, was the driving force that kept the German rocket development program moving forward.

Von Braun gave a technical lecture, which, onlookers noted in quick glance, seemed to bore the fuehrer. When von Braun was finished, a thick silence descended as he waited for Hitler to ask questions. Those present knew that it had been the fuehrer's habit, when shown a new-model tank or airplane, to ask questions about the smallest details for hours. Hitler remained mute.

Then the fuehrer and his entourage were escorted to a large clearing where two small test rockets, their snouts pointed upward, rested on launchers. Suddenly there was a belching of smoke and fire, a roar, and the rockets zoomed skyward and vanished into the low clouds. Hitler, his eyes protruding from beneath his visored military cap, watched expressionless—and said nothing.

Hitler was then shown a model of the huge A–3 rocket, so constructed as to reveal the internal mechanisms. Without doubt, the A–3 was the largest and most advanced missile in the world. It was 46.1 feet tall (the height of a five-story building), 65 inches in diameter, and more than 27,000 pounds. If and when the model went into production, it would be propelled by an enormous sea-level thrust of 56,000 pounds and would carry a 2,200-pound payload.

Knowing that a favorable nod from the fuehrer was crucial to continued rocket research and development, von Braun, Dornberger, and the other scientists and engineers were crestfallen over Hitler's seeming lack of interest. A short time later, the fuehrer, while picking at a salad and drinking mineral water at lunch in the mess hall, continued to keep his thoughts to himself. When he finally did say something about rockets, his remarks cast gloom over those at the table. Back in ''the early days,'' at the start of his rise to supreme power after World War I, the fuehrer said, he knew a man in Munich who experimented with rockets but he regarded him as a Spinner (crackpot). Only when he prepared to depart did Hitler utter a cautious opinion on the rocket demonstration: ''*Es war doch gewaltig*'' (It was impressive).[3]

Unknown to the scientists and engineers, the fuehrer was on the brink of taking an enormous gamble, one that might touch off World War II if the British and French contested his action. A few days after the Kummersdorf demonstration, on March 15, 1939, the *Wehrmacht* stormed into Czechoslovakia. The fuehrer announced that Czechoslovakia had become part of the Greater Reich, and he entered Prague as a conqueror.

Just over a year later, in April 1940, German forces invaded and rapidly overran Denmark and Norway; then Hitler's eyes turned to the West. At 11 A.M. on May 10, 1940, hundreds of panzers and scores of first-rate infantry divisions drove forward, and the blitzkrieg brought Belgium, the Netherlands, and France to their knees in an incredible six weeks.

Now Adolf Hitler, who had personally directed the lightning operation, was the overlord of most of Western Europe. Only defiant Great Britain, under newly appointed Prime Minister Winston S. Churchill, continued to resist. Convinced that England could be conquered by conventional weapons, Hitler struck the Peenemünde rocket research and development center off an earlier high-priority list.

In the Third Reich, Hitler's orders were law, and anyone who disputed them did so at the risk of his own life. However, Colonel Karl Emil Becker, head of the Army Weapons Office, who had been a science professor at the University of Berlin where he taught Wernher von Braun, was upset. In a face-to-face confrontation that turned into a virtual quarrel, Colonel Becker told the fuehrer that shutting down Peenemünde was a critical mistake. A few days later, an announcement was made in Berlin: Colonel Karl Emil Becker, rocket pioneer, had committed suicide.

Stubborn Great Britain, much to Adolf Hitler's dismay, proved to be a tough nut to crack. *Reichsmarschall* Herman Göring's vaunted Luftwaffe was battled to a draw by a relative handful of fighter pilots of the Royal Air Force, and a cross-Channel invasion by the German army was deferred.

Meanwhile, Colonel Walter Dornberger and General von Brauchitsch, the army commander, struck a secret deal that would allow rocket research and experiments to continue. Risking the fuehrer's anger (and possibly his own neck), von Brauchitsch authorized 5,000 specialists to be transferred from the army to Peenemünde. However, von Braun and the other scientists and engineers were handicapped in their work by a shortage of raw materials, instruments, and parts.[4]

Walter Dornberger was a professional soldier with a doctoral degree in engineering. Born in Giessen, a small university city near Frankfurt, where his father had been a pharmacist, Dornberger, now 43 years old, had fought as an artillery officer in World War I and was captured by the 2nd Division of the U.S. Marine Corps in October 1918. Handed over to the French, he spent two years in a prisoner-of-war camp, most of it in solitary confinement for repeated escape attempts.

Dornberger, a friendly, pleasant man, was fond of cigars and enjoyed off-duty bantering about eventual space travel. His unwrinkled, pink skin resulted from hideous burns he had received in 1932, when the powder in a rocket on which he was working exploded in his face. Doctors in a Berlin hospital said he would not survive. But for two years, he underwent extensive treatment and then returned to his rocket experiments.

Hitler's failure to crush the British Isles rapidly resulted in a new lease on life for Peenemünde. Secretly, the fuehrer had reached a momentous decision: he would launch his mighty war juggernaut against the Soviet Union, with which he had signed a "friendship pact" in 1939. Now Hitler remembered the new rocket weapons, and he granted Peenemünde the right of an SS-category quota on raw and manufactured materials—which meant top priority.

During late 1940 and early 1941, the Peenemünde team, under the overall supervision of Wernher von Braun, together with scientists and engineers at German universities and in industry, worked around the clock to develop the massive A–4 missile, an improved version of the earlier model in which Adolf Hitler had shown such remarkable disinterest at Kummersdorf West. Some A–4 components were in production, and the rocket's design was frozen.

With the *Oberkommando der Wehrmacht* (military high command) in Berlin putting the finishing touches on Operation Barbarossa (the invasion of Russia), the secret weapons experiments at Peenemünde were gaining widespread attention. Right next door to the Army Experimental Station, the Luftwaffe was working with equal haste and intensity to perfect the V–1 flying bomb. Both sides were competing vigorously for manpower, materials, and the favorable attention of the bigwigs in Berlin. Jealousy was rampant between the two competitors. There was no contact between the designers and engineers involved with the flying bomb and the missile, even though their experiments were closely related.

Both sides spied endlessly on one another, rejoicing in failures by the competing camp. Hitler was not yet sure which embryo *V* weapon, the flying bomb or the missile, should be given top priority, which of the two would be more effective in military operations.

During the first years of the war, the missile experiments were more advanced, and the designers of the flying bomb worked themselves into lathers as they peered across Usedom to watch the fiery tails of rockets streak into the Baltic sky. The Luftwaffe people had not yet been able to launch the pilotless airplanes, so they had to content themselves with registering repeated protests with Berlin that von Braun and his team were illegally hogging the air.[5]

At 3:00 A.M. on June 22, 1941, German armies of 3 million men swept across the Russian border on a 2,000-mile front, taking the Soviets by total surprise. Directing Operation Barbarossa at *Fuehrerauptquartier* in the Forest of Görlitz near Rastenburg in East Prussia, Adolf Hitler confidently expected to crush the Soviet army, capture Moscow, and end the campaign in three months–before the brutal Russian winter struck. Soviet defenses collapsed like houses of cards in a hurricane. In the first phase of Barbarossa, the Red Army lost perhaps 3 million men, 22,000 guns, and thousands of tanks and aircraft. However, the great ice clouds gathered behind the Aral Sea and sped across the Russian steppe with mounting violence and 50 degrees below zero temperatures. Five months after the Wehrmacht had plunged into the vastness of the Soviet Union, Hitler's freezing legions were bogged down before the gates of Moscow and Leningrad.[6]

For months, the fuehrer had been building up a head of steam in his hatred for the United States in general and President Franklin D. Roosevelt in particular. Hitler ranted about Roosevelt's personal attacks on him, and he was furious over America's lend-lease program with England.[7]

So on December 11, 1941, four days after the Japanese bombed the United States into war at Pearl Harbor, the fuehrer called his *Reichstag* (parliament) into session to rubber-stamp his declaration of war against the United States.

At Peenemünde, Wernher von Braun and Colonel Walter Dornberger were delighted to learn that they had a powerful friend within the fuehrer's inner circle: 37-year-old Albert Speer, an energetic type, bursting with ideas and innovations. Speer had been Hitler's favorite architect prior to the outbreak of the war in Europe. Hitler had commissioned the architect to draw up detailed plans and blueprints for a grandiose face-lifting of Berlin, a revitalization that would exceed by comparison the grandeur that had been ancient Rome. Speer's monumental building program had been set aside when the Third Reich went to war with the understanding that it was to be resumed when Germany triumphed.

Speer had been appointed to the post of minister for armaments and war production on February 8, 1942, when the current minister, Fritz Todt, was killed in an airplane crash. Speer was enthusiastic over the potential of remote-controlled missiles, and on June 13, 1942, he paid a visit to Peenemünde to witness a test firing of the A–4 (later to be dubbed the V–2). Before the visiting dignitary, in a clearing among the pines, towered a missile five stories high. Wernher von Braun and his staff were as full of suspense over the launching as was Speer, who knew what hopes the young inventor was placing on this experiment. For von Braun and his team, this was not the development of a weapon but a gigantic step into the future of space technology.

Wisps of vapor revealed that fuel tanks were being filled. Then, at the predetermined second, at first with a faltering motion and then with the roar of a supernatural unleashed giant, the rocket surged skyward and vanished into the low clouds.

Speer was thunderstruck at this technical miracle, at the way it seemed to abolish the laws of gravity, so that 13 tons could be hurled into the air. Wernher von Braun was beaming. All those present were jumping around and congratulating each other. Suddenly, they froze where they stood. A minute and a half after lift-off, a rapidly swelling howl indicated that the rocket was falling in the immediate vicinity. Speer, von Braun, and Dornberger remained standing—the others flopped to the ground. The A–4 smashed into the earth less than a half-mile away. Its guidance system had gone awry.[8]

Despite this apparent failure, von Braun and his rocket team were happy. The young scientist grinned and told Speer: "Well, at least we've solved the takeoff problem!"[9]

Two months later, another giant missile soared into the bright Baltic sky. It traveled a perfect course until, suddenly, the rocket bolted sideways and plunged thousands of feet into the sea.

Time was running out on the Peenemünde rocketeers. Their "spies" told them that the Luftwaffe engineers next door had nearly perfected the flying bomb, which could be built far cheaper (only $600 each) and much faster. However, Hitler had not yet made up his mind which revolutionary weapon he would put into mass production.

So on October 13, 1942, yet another A–4 missile was prepared for firing. Should this test fail, the fuehrer might well lose patience and cut off all rocket materials. A warning flare was sent up to alert those on Usedom. An engineer threw a switch, and a brilliant red-yellow flame shot from the rear of the rocket. Cables and wires fell away. With what sounded like an enormous clap of rolling thunder, the metal monster slowly lifted, then, with a strident screech, leaped into the cloudless sky.

Thousands of eyes were glued on the swiftly rising missile. One minute ticked past. The A–4 continued to climb. Two minutes. Four minutes. Von Braun stood as though mesmerized, then felt a surge of exultation: for the first time, a man-made object had broken through the sound barrier. Shouts of joy echoed across the launch area.

On the western side of Usedom, the Luftwaffe's flying-bomb camp was shrouded in gloom. The army's missile, after two previous test failures, had attained a speed of 3,500 miles per hour, soared nearly 35 miles high, and traveled approximately 120 miles.

That night, Colonel Dornberger gave a small party for Wernher von Braun and a few of the key rocket scientists. There was much to celebrate. Those present had been instrumental in creating a missile that for the first time reached the borders of outer space, proving that rocket power was practicable for interplanetary exploration.

Everyone present, in the words of a guest, "got a little tipsy." Walter Dornberger delivered a short speech of congratulation, concluding with: "Today, the space ship was born!" Raucous cheers erupted.

In the wake of the spectacular A–4 success, Dornberger, von Braun, and others on the rocket team felt that Hitler would now give the green light for vastly accelerated research and development. Their exuberance was soon snuffed out. There were only headaches. The needed supplies and technicians were slow in arriving, and Berlin bureaucrats increased their meddling into Peenemünde activities.

Dismayed, Walter Dornberger called on War Production Minister Albert Speer, an intelligent and enthusiastic booster of missile potential, and asked him to intervene with the fuehrer to energize the stalled A–4 program. On December 22, 1942, Speer approached Adolf Hitler at the Chancellery in Berlin and pleaded for approval to begin mass production of the A–4. Although the missile required extensive fine-tuning, Speer felt that matters could be rushed by getting the fuehrer to issue an order allowing mass production to commence.

Speer, as always, was eloquent and persuasive, but Hitler remained skeptical, filled with distrust of all innovations that went beyond the technical experience of the World War I generation. Reluctantly, however, Hitler signed the mass-production document.

In March 1943, at the urging of Colonel Dornberger, Albert Speer again prodded the fuehrer about hiking the priorities for A–4 development. The munitions minister reported back that Hitler was still not interested in giving his personal backing to the missile program. "The fuehrer," Speer said, "has dreamed that no A–4 will ever reach England." Dornberger turned the air blue with his curses, a dangerous activity in Nazi Germany where the fuehrer was concerned.[10]

In London, British intelligence had grown alarmed over bits and pieces of information indicating that the Germans were up to something big at isolated Peenemünde. Early in December 1942, MI–6 received a hand-written message, smuggled out of Europe, from a man who described himself as a Danish chemist working in Germany. He did not reveal his name. In his note, the Dane said that the Germans had recently fired a huge rocket that flew a distance of over 100 miles. He said that he had picked up sketchy details about the episode when he overheard two German engineers discussing it at a nearby table in a Berlin restaurant.[11]

A month later, Polish intelligence officers in London turned over to MI–6 a batch of astonishing materials that the underground in German-occupied Poland had smuggled into Switzerland and brought to London hidden in the handle of a courier's suitcase. There were rough sketches of the Peenemünde layout, the flying bombs, and the missiles.

Stewart Menzies, the chief of MI–6, and his top lieutenants were skeptical. It was felt that the "evidence" of unusual experiments being conducted at Peenemünde had been the result of Machiavellian intrigue by the Abwehr to get British intelligence to begin chasing its own tail and to create panic in official circles in London.

Two weeks after receiving the data from the Danish chemist, MI–6 received corroborating evidence from Pierre Ginter, a Luxembourg citizen who was one of thousands of foreigners sent to work at Peenemünde. Unlike the Pole and Slavic nationals, Luxembourg natives were considered to be *Reichsdeutschen*—pure Germans—so they received more privileges. Ginter was working as a night telephone operator, and in some manner he managed to steal partial drawings of the A–4 missile and the flying bomb. Permitted to go home on a short leave, Ginter, at the risk of his life, took the supersecret drawings with him and turned them over to the underground cell *Famille Martin* in northern France, from where the documents were sneaked across the Channel to England.

Up to this point, the collective evidence about German missile development was inconclusive and conjectural. Then, from an unexpected source—the Germans themselves—came chilling confirmation.

During fighting in North Africa in early 1943, the Allies captured two Wehrmacht generals, Wilhelm von Thoma and Ludwig Cruewell. They were taken to England and put in a room that British intelligence had bugged. It was hoped that the two generals, who had long known each other and would have no qualms about expressing their views, would drop clues about future German military plans.

A written transcript of the Germans' long and often rambling conversations was sent to Reginald Jones, MI–6's young scientist. Jones, in turn, passed it along to his number-two man, Charles Frank, who focused on a few eye-opening remarks. General von Thoma had told his fellow prisoner of war (POW) about General Walther von Brauchitsch (the commander of the German army) taking him to Peenemünde a few months earlier to witness the firing of a monstrous rocket. Von Brauchitsch, von Thoma added, had confided in him that the missile would be ready for tactical deployment within one year.[12]

Late in June 1943, Royal Air Force (RAF) Flight Officer Constance Babington-Smith, regarded as an expert among experts in photo interpretation, was seated at her desk at the British Central Intelligence Unit (CIU) in the sleepy village of Medmenha, 30 miles west of London. The function of the CIU was to interpret aerial photographs.

Babington-Smith's sharp eyes, peering through a magnifying glass, focused on Peenemünde region photos that had just arrived. British photo reconnaissance planes had swept over the barren region (as well as much of the north German Baltic coast so as not to arouse suspicions). She picked out an elongated shape on one of the prints. New photos came in the next day; other similar shapes were detected. Babington-Smith and her colleagues concluded, without a doubt, that these cigar-shaped objects were huge missiles.

Combined with earlier bits and pieces of related evidence, British intelligence concluded that the Germans had rockets of enormous destructive power and long range and that Hitler had already set October 1943 as the date for deluging London with explosives. Only a few technical problems might postpone the launchings.

On the morning of July 7, 1943, a black-bodied Heinkel was winging through a murky sky bound from Peenemünde to Adolf Hitler's battle headquarters, *Wolfsschanze* (Wolf's Lair), in East Prussia, behind the flaming Russian battlefront. On board were Wernher von Braun, Colonel Walter Dornberger, and 35-year-old Ernst A. Steinhoff, technical director of the rocket team. They had been summoned by the fuehrer, who was eager to learn details of the progress of the missile development project.

Arriving at Wolf's Lair, the three rocketeers were met by Albert Speer, the armaments boss, and the four men had to cool their heels for 24 hours; the fuehrer was too busy to see them. In fact, he had been so occupied directing his armies in warding off the Russian hordes and swarms of tanks that he never found time to visit Peenemünde. So the three scientists had brought with them a color film of a perfect rocket takeoff as well as scale models with which to conduct a briefing.

Along with Adolf Hitler, a glittering lineup of German brass gathered in the Wolf's Lair theater: Field Marshal Wilhelm Keitel, chief of the Oberkommando der Wehrmacht and the fuehrer's most trusted military advisor; General Alfred Jodl, also a close aide; and General Walter Buhle, chief of staff of the German high command.

Wernher von Braun was stunned by Hitler's physical appearance. This was not the same vibrant man who had visited the demonstration of rocket firing at Kummersdorf West in May 1939. Rather, burdened by one disaster after another on battlefields in the Soviet Union and in North Africa, the fuehrer was a weary, slightly stooped figure who looked far older than his 54 years.

The theater lights dimmed. Filmed with Teutonic expertise, the moving documentary was a dramatic, color presentation of a long, shiny missile being hoisted onto the firing table by a *Meillerwagen* (a long, flatbed vehicle) with its hydraulic boom. All the while, von Braun delivered a running commentary.

Unlike his indifferent attitude at Kummersdorf West four years earlier, the fuehrer gawked as though mesmerized by the unfolding scenario. Then the highlight of the film: the A–4 missile made its magnificent ascent, trailing a fiery plume, until it disappeared into a bank of billowing white clouds.

At the end, these words were flashed on the screen: "We made it after all!"

Then the lights were flipped on. No one spoke. All eyes were on the fuehrer. He sat motionless and silent, as though deeply emerged in thought. Suddenly, Hitler leaped to his feet, grabbed von Braun and Dornberger by the hands and pumped them vigorously, in an uncharacteristic display of emotion. Hitler even seemed to apologize for his earlier skeptical attitude.

"I thank you," he told the rocketeers. "Why was it I could not believe in the success of your work? If I had had these rockets in 1939, we should never have had this war. . . . Europe and the world will be too small from now on to contain a war. With such weapons humanity will be unable to endure it!"[13]

As the Peenemünde contingent prepared to leave, it was announced that the rocket project would be given top priority and the A–4 would be renamed V–2 (*V* for *Vergeltungswaffe*, or vengeance weapon).

Speer accompanied the fuehrer back to his command bunker. His imagination kindled by the V–2 film, Hitler declared: "Speer, the missile is a measure that can decide this war. This will be the decisive weapon of the war!"[14]

Time and again in the conversation, Hitler remarked about Wernher von Braun. "Weren't you mistaken?" he asked. "You say this young man is 31? I would have thought him even younger." The fuehrer thought it astonishing that so young a man could already have helped to bring about a technical breakthrough that would change the face of the future.[15]

Wernher von Braun returned to Peenemünde with mixed emotions. He was elated that the fuehrer had so enthusiastically heaped praise upon him and the rocket program. On the other hand, he had always envisioned the climactic product of his life's work to be the creation of a vehicle in which man could peacefully explore the universe. A few days later, von Braun was notified that the fuehrer had granted him a titular professorship, an extremely high honor in Germany, one that could be granted only by the head of state, a tradition going back hundreds of years.

# 4

# A Bitter Clash of Viewpoints

A thick veil of tension hovered over the conference room at London's Whitehall, the nerve center of Great Britain's war effort, as Prime Minister Winston Churchill convened an emergency session of his War Cabinet. Among those present were Deputy Prime Minister Clement Attlee, Foreign Secretary Anthony Eden, Home Secretary Herbert Morrison, Minister for War Production Lord Beaverbrook, General Hastings ''Pug'' Ismay, and Churchill's scientific advisor Lord Cherwell. It was June 28, 1943.

For months, those in the upper councils of the British government, scientists, intelligence chiefs, and military leaders had been disagreeing violently among themselves as to whether Germany had actually developed a rocket. Most of the scientific experts, led by Lord Cherwell, refused to concede that the Germans were years ahead of the British not only in missile development but in other crucial areas. Cherwell and other dissenters no doubt would have been shocked had they known that a massive V–2 missile already had been launched successfully at Peenemünde on October 3, 1942.

Lord Cherwell, arrogant, irascible, and brilliant, brushed off the mass of accumulated intelligence that indicated a German rocket had been developed as a devious deception plot by the Abwehr. Now at the somber Whitehall conference, Cherwell (his real name was Frederick Lindemann) reiterated his long-held view that a large, long-range rocket was a technical impossibility.

On the following night, high-ranking politicians, military chiefs, scientists, and experts of numerous stripes met in the underground War Cabinet room for a showdown over the conflicting evidence. Leading the discussion was 36-year-old Duncan Sandys.

Sandys had been badly injured in an automobile accident while posted to Britain's rocket station in North Wales two years earlier and had retired from active duty. In April 1943, at the recommendation of the Imperial General Staff (IGS), Churchill had appointed the energetic, keen-witted Sandys as the coordinator of information on the German rocket threat. So important was this position that Sandys reported directly to the War Cabinet, not through customary channels.

In singling out Duncan Sandys, the Imperial General Staff noted that he was eminently qualified. What was crucial, the IGS stressed, was that the person named to the post approach his task with an open mind and not be the victim of preconceived notions or vested interests, as might be the case if a career intelligence officer or a scientist were named. Sandys' function was code-named Bodyline.

Sandys' appointment to the prestigious and influential post immediately brought brickbats flying in his direction from scientists and other technical experts who had been involved in analyzing the German missile program, including the charge that Sandys had been named solely because he was the son-in-law of Winston Churchill. What was more, they complained, Sandys was not a scientist but a politician and an army officer.

If most of the British scientists were unhappy, Lord Cherwell was downright furious. A professor of physics at Oxford University, Cherwell was in his mid–50s, had a towering opinion of himself, and had a reputation for injecting himself in all aspects of the government. Back in 1941, when the fledgling Royal Air Force Bomber Command was suffering heavy losses in raids over Germany, Cherwell independently organized an investigation of Bomber Command's techniques and effectiveness. Cherwell posed the question: Were the almost nightly RAF raids hitting their targets? Cherwell named D. M. Butts of the War Cabinet secretariat to conduct the research on the RAF bombing survey, which was based on more than 600 photographs taken on some 100 bombing missions, in which crews returned to claim success.

In August 1941, Lord Cherwell (smugly perhaps) turned over to Churchill the Butts analysis, which indicated only 10 percent of the RAF bombs were falling on or near the targets. The prime minister forwarded the study to Air Chief Marshal Sir Charles Portal, chief of the Air Staff, with a terse note scrawled on it: "This is a very serious paper and seems to require your urgent attention."[1]

Portal and his staff were furious at Cherwell and rejected outright the conclusions in the Butts study. Churchill's scientific advisor won no friends within the RAF.

Now, three years later, Cherwell's deep resentment toward the likable Duncan Sandys resulted from hurt pride and jealousy over the appointment of a much younger, "totally unqualified" man to such an important post. Consequently, this personal envy may have blinded Cherwell to the German missile threat.[2]

Sandys opened the Whitehall conference by showing aerial reconnaissance photos of Peenemünde. Starkly evident were several white rockets. Then Sandys made a one-hour presentation of the total intelligence picture, concluding with an urgent recommendation that Peenemünde be bombed "at the earliest possible date."

Although most of those present were impressed by the thoroughness and the keen insight of Sandys' briefing, skepticism remained that the Germans even had a rocket. Home Secretary Morrison ventured the view that the flood of intelligence reports and the ease with which the aerial photographs of Peenemünde were obtained made the entire matter suspicious. "This is, after all, supposed to be a closely guarded war secret," Morrison declared.

Lord Cherwell then got in his licks. The materials Sandys had assembled, he pointed out, merely proved that the Germans quite cleverly had planted them in order to divert the British into chasing off in all directions. Had the rockets been painted white so they could be easily spotted from the air? It was crucial to avoid stumbling into a cunningly laid German trap, Cherwell snorted.

Cherwell's staunch opinion that the Germans did not even have long-range missiles in their arsenal was backed by Alwyn Crow, who was widely recognized as one of Britain's foremost rocket experts. Back in 1936, Crow had been designated to develop an antiaircraft rocket, and four years later, with Britain at war, he had produced short-range, two- to five-inch rockets, powered by solid cordite fuel.

Both Crow and Cherwell had dismissed the possibility of developing the British cordite rocket to carry a one-ton warhead for a distance of over 100 miles. Any such rocket would be a monster, weighing at least 75 tons, and could not be launched, they concluded.

Back in August 1942, Crow had visited the laboratory of Isaac Lubbock, an engineer of modest reputation within the British scientific community, who was performing pioneer research into liquid fuel for rockets. Missiles powered by liquid fuel would dramatically reduce the weight of long-range rockets using solid cordite fuel. Crow viewed a successful demonstration of Lubbock's liquid-fuel rocket.

Crow was so enamored of his own solid-fuel project that he chose to ignore Lubbock's promising new development. Crow, as with Lord Cherwell, adamantly refused to acknowledge that a practical long-range missile powered by liquid fuel could be developed, for no scientist or engineer in the Allied world had designed a pump that could supply fuel quickly enough to the rocket's combustion chamber. Both Crow and Cherwell were unaware that a German engineer, Walter Thiel, already had designed and perfected a revolutionary turbopump driven by a gas turbine.

Now, at the June 1943 conference, Winston Churchill, puffing on a long black cigar, called on young Reginald Jones, a former pupil of Cherwell at Oxford. Jones felt ill at ease. All eyes were on him, and everyone in the smoke-filled underground room was his senior by far in age, experience, and in every other way. The whole affair had a sense of intense confrontation.[3]

Cherwell listened stonily as Jones began, point by point, to demolish his former professor's arguments. Jones declared that Peenemünde was one of the Third Reich's most critical research centers, and the voluminous intelligence gathered from a variety of sources proved beyond the shadow of a doubt that the Germans had long-range rockets of enormous power and destructive capability.

Intelligence reports indicated, Jones added, that Adolf Hitler already had set a date for bombarding London with missiles and that the need for German rocket scientists and engineers to smooth out technical dents had postponed the launchings.

All the while Jones was speaking, Cherwell was growing hotter under the collar. His mood was not improved when his old friend Prime Minister Churchill kept looking at him and nodding affirmatively at key points in Jones' presentation.

Duncan Sandys and Reginald Jones, the two youngest men in the room, carried the day. After prolonged and often raucous debate, Churchill and the War Cabinet agreed that Peenemünde should be heavily bombed at the earliest time permitted by weather conditions.

Across the Atlantic Ocean, the U.S. military establishment was paying scant heed to the alarming reports coming from intelligence sources in London concerning formidable evidence that the Germans had developed a long-range rocket. One of those who did absorb the reports was Theodore von Karman, who taught at the California Institute of Technology in Pasadena.

Hungarian-born von Karman had been a U.S. rocket pioneer. Back in the early 1930s, he collected a few engineers and scientists to study the possibility of a liquid-fueled rocket for space exploration. The group met with some successes, but von Karman failed in repeated attempts to interest the U.S. War Department in funding his project. Privately, the generals in Washington, schooled on the static warfare tactics and conventional artillery of World War I, sneered at the rocketeers at Cal Tech.

What was more, the United States was bogged down in the throes of a stultifying economic depression, and the nation's mood after World War I virtually had been to disarm and stay out of other nation's quarrels. So there had been no funds to energize von Karman's Jet Propulsion Research Project, even had the Washington brass approved.

After the woefully unprepared United States was bombed into global war on December 7, 1941, von Karman was able to interest the War Department in the potential of rockets as military weapons. But the amount of money dished out for what von Karman now called the National Defense Research Committee was puny.

In the months ahead von Karman periodically made pleas to Washington for more financial aid, pointing out that intelligence reports coming out of London indicated that the Germans were pouring hundreds of millions of marks into the development of a long-range missile and that scientists in the Third Reich appeared to be 20 to 25 years ahead of the Americans and the British in rocket technology.

One of the handful in Washington who paid heed to von Karman's warning was Colonel Gervais Trichel, the energetic chief of the rocket branch of Army Ordnance. In July 1943, Trichel, after reading von Karman's report on the advanced status of German rocketry, urged the Cal Tech scientist to organize the American equivalent of the German V–2 program. Such an ambitious goal required heavy funding by Washington, so missile research and development in the United States continued to lag far behind that of the Third Reich.

While Theodore von Karman in the United States was begging Washington for rocket development funds, alarming evidence that the German long-range missile threat was real continued to reach London. On the moonless night of August 3, 1943, a British undercover agent was walking briskly toward a clandestine rendezvous with a German spy on a dark corner in neutral Lisbon, Portugal, a hotbed of cloak-and-dagger intrigue. Arriving at the designated spot, the Brit reached out and took a large envelope handed to him by the German. Then both shadowy figures walked off in opposite directions.

The documents in the envelope were rushed to MI–6 in London, where they became known as the Lisbon Report. It may have been the most devastating betrayal suffered by the fuehrer during the war.

Among a mass of technical detail, the Lisbon Report disclosed information on Peenemünde rocket research and development. Stewart Menzies was convinced that Admiral Wilhelm Canaris, Hitler's devious spymaster, was behind the Lisbon Report. Who else in Germany could have collected such a mass of detailed, apparently authentic and sophisticated data about a supersecret project without attracting suspicion, then have the means for getting the materials into the hands of the British?[4]

Almost from the hour Adolf Hitler's booted legions had conquered Denmark, Norway, Holland, Belgium, Luxembourg, and France in the early 1940s, underground networks had been feeding highly valuable information to Allied intelligence in London. Also active in clandestine surveillance of German activities in northern France were British intelligence "sleepers," who had been living routine lives for years, and spies involved with Polish intelligence (codenamed F2) and the underground Polish Continental Action (*Akcja Kontynentalna*).

It was risky business. Striking periodically with great precision, the Gestapo arrested and executed hundreds of men and women. Many spy networks arduously built up over the years were systematically destroyed.

Beginning in July 1943, scores of underground agents in northern France, acting in tiny groups or solo, began sending London scraps of information about curious diggings that the Germans were engaged in. These excavations, scattered along the English Channel coast for 150 miles, from Cherbourg on the west to the Pas de Calais on the east, appeared to be installations for firing at London. But firing what?

Alarmed British intelligence dispatched RAF reconnaissance planes to swoop low all along the Channel coast, and thousands of photographs were taken. The prints disclosed the widespread diggings.

Unaware of precisely what the excavations were intended to accomplish, Allied intelligence began to float conjectures. These strange facilities were refrigerating apparatuses and were designed to cripple Allied bomber fleets by dropping ice clouds over England, one rumor had it. Another held that the Germans were going to fling across the Channel huge containers of poison gas that would wipe out every inhabitant in the British Isles. Yet another rumor reported that these containers would land in England and spread the ''Red Death,'' although no one could define exactly what this plague would be. Red Death sounded ominous, however, so it gained considerable currency.

In northern France on the hot, sultry afternoon of August 12, 1943, an agent of the *Réseau Agir* (Network for Action) was sipping wine in a cafe in the Seine River port of Rouen, when he overheard two building contractors discussing mysterious construction work being carried out by the Germans. The agent reported the conversation to the chief of Réseau Agir, 45-year-old Michel Hollard.

Hollard, short and stocky, had become a spy on his own initiative when the Wehrmacht marched into Paris three years earlier. He quit his job as an industrial designer and obtained a position as a salesman for charcoal-burning gas generators for automobiles. This job permitted him to roam France and later would explain his presence near the Swiss border, which by mid-1943, he had crossed 43 times with high-grade intelligence for the Allies.

On the basis of the report from the Rouen cafe, Hollard strolled into the Rouen employment office the next day and, clad entirely in black, announced that he represented a religious organization interested in the spiritual uplifting of laboring men. Clutching a Bible, he asked if there happened to be any building projects taking place in the area. ''Why, yes,'' he was told. ''There's one at Auffay [20 miles from Rouen].''

Two hours later, the masterspy was at Auffay, wearing workman's clothes. Strolling around, he stumbled onto a large clearing where a few hundred laborers were pouring concrete. Hollard grabbed a wheelbarrow, filled it with bricks, and pitched in. No one stopped or even challenged him. Most of the workers were foreign, but one who could speak French told Hollard they were building garages. Hollard's curiosity deepened. Garages, in the middle of a forest 20 miles from the nearest city?

Hollard's keen eye was attracted to a 50-yard strip of concrete that appeared to be some sort of a ramp. Furtively, he removed his compass and found that a guideline of string laid along the ramp was aimed in the direction of London. Aware only that a mysterious ''crash'' project was being built by the Germans, Hollard rushed a report to England. It exploded like a blockbuster bomb in the highest councils of the government and military.

What Hollard did not know was that Allied leaders were deeply worried over repeated clues that ''flying bombs'' were about to be showered on London. Hollard was instructed to drop everything else and get more details.

The Réseau leader and four of his men began crisscrossing the Channel coastline on bicycles, and they discovered in less than one month 60 identical mystery ramps. They kept pedaling and by mid-November had detected 40 more, all located in a band 20 miles deep and all aimed at London.

What were these strange ramps? Hollard was ordered to find out. He placed a volunteer (code-named André) with the labor force, and he was assigned to a desk job in the Bois Carré, where more of the strange construction was taking place. A week later, Hollard's spy reported to his chief that the German in charge at the site kept a master plan in an inside pocket of his gray-green overcoat and that he wore the long garment at all times, even in his office. André noticed that the only time the German removed the overcoat was at about 9:00 A.M., when he went down the hall to heed the call of nature.

For a week, André timed the absences. They lasted between three and five minutes. On the eighth day, the project manager shed his coat and left to perform his morning ritual. Like a jungle cat on the prowl, André stole into the German's office, removed the master plan, and made a rapid tracing. Just as the Frenchman had replaced the blueprint in the overcoat and left the room, the German returned—unaware that he had just been the victim of a monumental intelligence coup.

Two days later, Michel Hollard, evading the German patrols that roamed the region, slipped across the Swiss border. One false move would mean his death. He was dressed like a woodcutter, carried an axe in one hand, and had a sack of potatoes slung over a shoulder. Hidden among the potatoes was the blueprint tract of the German mystery site.

Within 48 hours, André's handiwork reached London. Allied leaders were shaken: here was a drawing of one of the scores of flying-bomb launching bases from which Hitler planned to demolish London.[5]

At Peenemünde West, where Luftwaffe scientists and engineers were conducting tests, a Heinkel 111 carrying a V–1 flying bomb lifted off and at a designated point released the weapon to fly on alone. Its steering mechanism went awry; it flew much farther than planned and landed in a tulip field on the island of Bornholm, off the coast of Denmark, and failed to explode.

During the German occupation, Danish police continued to work on civilian criminal cases, but many officers were also in contact with the underground. By happenstance, Police Inspector John Hansen was nearby when the flying bomb landed. Hansen did not know what the strange-looking object was, but he noted the German markings. Along with a friend, Danish sea captain Hasager Christiansen, Hansen quickly snapped some photographs and concealed parts of the bomb.

An hour later, a German security contingent reached the scene and took charge of what remained of the V–1. Hansen, meanwhile, drew a detailed sketch of the object based on his memory and the pieces in his possession.

Eight copies of a report, which included sketches and photographs, were compiled and sent by eight Danish couriers, each taking a different route, to

England. It was hoped that at least one would make it through the tight German security dragnet. One courier was apprehended, a few never got out of Denmark, but one arrived in London a few days later. The detailed report was carried by Major V.L.U. Glyth of Danish intelligence who, with the Gestapo on his tail, fled into Sweden just before his pursuers caught up with him.[6]

This had been the first flying bomb to fall outside of Germany, and most British scientists were elated to get such a wealth of eyewitness information. From this material, they could construct a prototype with a view toward designing a defense against the V–1.

Lord Cherwell, Winston Churchill's scientific advisor, ventured the opinion that the warhead of the flying bomb, based on the information received from the Danish underground, would weigh no more than 1,100 pounds. Therefore, he concluded, the flying bomb was of no real danger to England. Besides, he conjectured, this V–1 had been but a test model and the weapons were not being mass-produced.

# 5

# Operation Hydra

Purple shades of dusk were cloaking England's midlands at 7:57 P.M. on August 17, 1943, when RAF Wing Commander John H. Searby and his six-man crew lifted off in their four-engine Lancaster. Winging eastward behind them was a bomber force of 433 heavies (Stirlings, Halifaxes, and Lancasters) and 65 Pathfinders. Target: the German experimental rocket center at Peenemünde. Operational code name: Hydra.

So crucial was the mission that Winston Churchill had approved risking nearly the entire British bomber fleet in order to wipe out the target. Prior to takeoff, Air Chief Marshal Sir Arthur Harris briefed 4,000 airmen not just to bomb the facilities but to kill the scientists and engineers. For security reasons, crews were told that special radar equipment was being made there.

As part of the RAF deception, eight fast Mosquito bombers flew toward Berlin, 90 miles south of Peenemünde. At a designated point near the German capital, the Mosquitoes began releasing a batch of 2,000 aluminum foil strips at one-minute intervals. The strips (known to the British as Window and to the Americans as Chaff) were a revolutionary new scheme in electronic warfare.

Two British scientists, Robert Cockburn and Joan Curran, had been laboring for more than a year to perfect this ECM (electronic countermeasure). They had found that if strips of foil were dropped from airplanes in large quantities, a devastating amount of confusion would be inflicted upon German air defenses. Window would black out enemy radar screens, snarl direction-finding equipment, and create so many false "echoes" that bewildered German radar operators on the ground would gain the impression that hundreds, or even thousands, of bombers were attacking.

At 11:05 P.M., the eight Mosquitoes were over Berlin and began dropping flares and bombs, giving the impression that they were the main force. At the same time, the RAF heavies were winging toward Peenemünde. German air defenses were thrown into mass confusion. In Holland, Luftwaffe General Joseph

Kammhuber, commander of the Twelfth Air Corps, was cut off from his airfields dotting northern Germany. Mysteriously, Kammhuber's main radar and communications center at Arnhem, where two Luftwaffe technicians were British spies, suddenly ceased operating.

Clearly, a large Allied bomber stream was over Germany, but no one had the means to determine its destination. Finally, the order went out: "All night fighters head for Berlin!" Two hundred of them rushed to the capital and chased around the moonlit sky in search of enemy bombers that were not there. The seven Mosquito decoys were already high-tailing it for home (one craft had been shot down).

Just past midnight, Wing Commander John Searby was circling high above Peenemünde in his Lancaster. His job was to direct the approaching force of 597 heavy bombers to the targets by radio telephone.

Down below, Colonel Walter Dornberger was awakened by a terrific blast. Glass from shattered windowpanes covered his bed. Dashing to look outside, he could see that Peenemünde was bathed in brilliant iridescence from flares dropped to light the targets. Throwing his tunic over his pajamas, he walked outside.

From high in the sky came the dull, sing-song drones of swarms of powerful airplane engines. Scores of ack-ack guns opened fire at the intruders. Eerie rustling sounds told of clusters of falling bombs. Mighty explosions rocked the vicinity.

"Where in the hell are our night-fighters?" Dornberger reflected bitterly as he made a dash the short distance to the concrete command bunker. While crouched and running, the colonel spotted Wernher von Braun, his hair gray with ashes, standing in front of the bunker, peering skyward. A nearby exploding bomb sent Dornberger and von Braun hustling into a long, brightly lit room crammed with grim-faced people.

Dornberger had no intention of staying put, despite the rain of bombs. He began barking out orders. Two officers were directed to round up a labor gang and start fighting the fires. Von Braun was told to rush to the construction bureau, collect as many drawings and records as he could, and stash them in a safe place.

Dornberger dashed out of the bunker on von Braun's heels into an inferno of fires, explosions, ack-ack guns, crumbling walls, and roaring bomber engines. Thick clouds of smoke covered much of the installation. People, their faces blackened by soot, charged about helter-skelter. Four thousand technicians and their families lived in what was called "the Settlement." There one of the first bombs scored a direct hit on the house where Walter Thiel, who was in charge of rocket propulsion and who had developed the A–4 engine, his wife, and four children were asleep.

Incendiaries created havoc in the Trassenheide camp where thousands of Russian prisoners of war and conscripted Polish laborers slept. Hundreds of shacks caught fire. Panic-stricken men raced about, many of them flaming torches. They tried to claw under the barbed-wire enclosures with their bare hands but were forced back by machine-gun fire from the guards. Other frantic laborers were turned back by snarling Doberman pinschers on the other side of the fence.

Finally, the constant roar of airplane engines faded into the distance, after 1,600 tons of high explosives and 280 tons of incendiaries had been dropped. The Royal Air Force had paid a heavy price in blood. On the return trip to England, the slow-moving bomber fleet, with only its own guns for protection, was pounced on by swarms of swift Luftwaffe night-fighters. In the bright moonlight, it turned into a turkey shoot. Forty bombers were shot down, along with 240 airmen.

At dawn, heartsick Wernher von Braun and Walter Dornberger located a light airplane that had escaped the carnage and flew over Peenemünde to assess the damage. The destruction was heavy, but by no means had the experimental center been devastated beyond partial repair.

The two men noted a peculiar fact: the flying-bomb side of Usedom island had been virtually untouched. It was of no consolation to the army missile men that the British obviously had considered their own operation to be far more important than the buzz-bomb work of their despised Luftwaffe rivals next door.

In the wake of the bombing, the Luftwaffe chief of staff, General Hans Jeschonnek, was blistered over the telephone by his boss, Reichsmarschal Hermann Göring for the air-defense confusion. Göring, the Luftwaffe commander, needed a scapegoat and Jeschonnek, a courageous and capable officer, was it. Göring had no way of knowing that the real culprits were a pair of unheralded British scientists, Curran and Cockburn, who had conceived the Window radar-jamming technique.

The heavy-jowled reichsmarschal charged General Jeschonnek with conducting himself "like a recruit." Two hours later, at 9:00 A.M., Jeschonnek was found dead on the floor of his Berlin office, a pistol in his hand. Nearby was a note: "I cannot work with Göring any more. Long live the Fuehrer."[1]

On a beautiful Saturday morning, August 21, a funeral Mass was held at Peenemünde, presided over by a Catholic priest and a Lutheran minister. Seven hundred and thirty-five charred corpses were buried in a common grave. Most of those killed in the bombing were laborers from Trassenheide; 178 German technicians lost their lives in the Settlement.

At Wolf's Lair on August 22, a somber Adolf Hitler was discussing the Peenemünde disaster with Albert Speer; SS Reichsfuehrer Heinrich Himmler; and Himmler's protégé, 40-year-old SS *Brigadefuehrer* (Brigadier General) Hans Kammler, a scientist who had the traits his boss demanded—a keen brain and total lack of scruples. Himmler, forever lusting for greater personal power, had a simple proposal for the fuehrer: the missile program had been threatened because someone betrayed Peenemünde; therefore, the entire rocket organization should be put in Himmler's hands.

Just before dawn, when it was Hitler's customary time to go to bed, he reached a number of crucial decisions. Everything connected with missile research and development would be taken from army jurisdiction and put under the control of Reichsfuehrer Himmler. Actual production of V–2s would remain under the direct supervision of Albert Speer. V–2s—at the rate of 30 per day—would be built in mammoth underground factories in the rugged Harz Mountains, about 125 miles southwest of Berlin. Missile-firing tests would be shifted to Poland, out of range of Allied bomber fleets. The V–2 side of Peenemünde would be rebuilt partially and von Braun and the other rocket scientists and engineers would continue to conduct research there.

Reconstruction work at Peenemünde began with typical Teutonic energy and resourcefulness. Repairs would be made only on essential buildings, and research activity would be kept inside during the day when Allied photo reconnaissance planes might be overhead. Everything was to be done to give the illusion that Peenemünde had been destroyed and abandoned. The huge bomb holes were left intact and the rubble and debris were shoveled into open spaces where it would be picked up by air photos. Scores, perhaps hundreds, of burned houses in the Settlement and in Trassenheide were left untouched, and the technicians, their families, and surviving slave laborers were scattered into villages in the region.

Wernher von Braun's mad dash during the height of the bombing had saved most of the crucial blueprints and technical documents. Within an incredibly short period of time, research to increase the V–2 range and accuracy and to develop even longer range missiles was in full swing at Peenemünde.

In London, Air Chief Marshal Charles Portal, chief of the Air Staff, was among those highly optimistic about the success of Operation Hydra. Photo interpretation and other evidence had convinced Portal that Peenemünde had been "knocked out for at least six months."

Although Heinrich Himmler had elbowed his way into control of the V–2 research, he now set his sights on taking over the remaining component of the missile program—production. It had been his pattern to confer an honorary rank in the SS upon almost every government official whose clout he sought to control or influence. So he called in Albert Speer and told him that he had reserved a particularly high distinction for him; the armaments boss would be made an honorary SS *Oberstgruppenfuehrer*, a rank equivalent to that of a full general in the German army.

Aware of Himmler's motive and reputation for being a master conniver, Speer declined the offer with polite phrases, pointing out that the army also had offered him high titular ranks and that he would offend the army generals if he accepted the SS honor.[2]

It was not just Winston Churchill, his War Cabinet, and the Imperial General Staff that were elated over the apparent destruction of the German V–2 project at Peenemünde. Reichsmarshal Hermann Göring, the rotund Luftwaffe leader, was also delighted with the apparent demise of the rival army long-range missile program.

Göring and Heinrich Himmler had long been bitter rivals, each seeking ever wider power. Now that Peenemünde had been heavily damaged, Göring detected the opportunity to leap past Himmler in the ongoing game of one-upsmanship. The Luftwaffe chief suggested to the fuehrer that top priority should now be given to the V–1 (flying bomb) because it had about the same range, could carry as big a warhead, and could be mass-produced much quicker and for one-tenth the cost of the V–2.

Hitler bought Göring's sales pitch, and plans were made to produce 50,000 of the winged robots, most of which would be launched against London. The devious devices had speeds of 440 miles per hour (much faster than any Allied warplane could travel) and had a timing apparatus that cut off the engine over the target. When the plummeting robot hit the ground or a building, it would explode with the impact of a 4,000-pound blockbuster bomb.

Across the English Channel in London on August 27, 1943, the Secret Intelligence Service (SIS) convened to review the mountain of materials collected by Duncan Sandys' Bodyline. Sandys' main sources had been an agent working inside the Ministry of Armaments in Berlin, aerial photographs, POW interrogations, bits and pieces from scores of spies and underground agents in Germany and countries occupied by the Wehrmacht, and radio intercepts.

After intense study, the SIS reached the conclusion that the Germans at Peenemünde were developing not just the long-range missile but also a pilotless airplane filled with explosives (the V–1). Lord Cherwell and Alwyn Crow read the SIS conclusions but remained unconvinced. Crow brushed off the overwhelming evidence: "We are of the opinion that the possibility of such a development in Germany can be ruled out."

Rather than bring cohesion among British scientists, the SIS report touched off a month of vitriolic debates. At one tension-racked conference, Crow shrugged off the most recent air reconnaissance photos of two V–2 missiles at Peenemünde as inflated barrage balloons. Then Colonel Kenneth Post, Duncan Sandys' number-two man, asked evenly, "Then why are balloons transported on heavy railroad flatcars?" Crow's face flushed crimson, but he remained silent.

In mid-October 1943, scientist Isaac Lubbock returned from a trip to the United States to reveal that American engineers had developed a gas-turbine pump, such as the kind that would be required to provide liquid fuel to a huge missile, thereby vastly lowering its weight. At a Defense Committee conference, chaired by Prime Minister Churchill, to consider new evidence in the heated rocket controversy, Lubbock presented an accurate sketch of a German rocket based on its use of liquid fuel. Such an alleged rocket was technically impossible, Crow bellowed.

Lord Cherwell also was irate over Lubbock's sketch and theory. He lambasted Lubbock as a "third-rate engineer" and charged him with "meddling" in areas about which he knew virtually nothing. Unaware that Hitler had ordered a higher priority for the flying bomb over the V–2 missile, Cherwell and Crow argued:

"If the Germans have long-range missiles, why haven't they launched attacks against London with them?"

Across the Atlantic Ocean in December 1943, American scientists leaped into the raging controversy that until now had been all British. Vannevar Bush, whom President Franklin D. Roosevelt had selected to head the Office of Scientific Research and Development, and other American experts also doubted the feasibility of a huge, long-range missile carrying a one-ton warhead.

After the saturation bombing of Peenemünde, British intelligence was alert to the possibility that Hitler might resume missile testing at some other remote locale within what the fuehrer called the Greater Reich. To eavesdrop on Hitler's councils of war and possibly to pick up a clue as to German plans for future V-2 testing, the British utilized one of history's most closely guarded secrets (code-named Ultra), an ingenious device that intercepted and deciphered top-secret, encoded German military and government wireless messages.[3]

Ultra had its origin five years earlier. Before Hitler invaded Poland in September 1939, the Wehrmacht adopted Enigma, an encoding machine whose ciphers the Germans considered to be unbreakable. Even if an enemy were to steal or capture an Enigma machine, the Oberkommando der Wehrmacht was convinced, it would be of no use without the knowledge of the keying procedures, which were changed almost daily.

In 1938, Polish intelligence, which was working with MI-6 against the Third Reich, pulled off a spectacular coup: one of its agents stole an Enigma right out from under the noses of the Germans. Shipped to London, the device was studied intently by a group of leading British scientists, mathematicians, and cryptanalysts, who concluded that there was but one way to penetrate Enigma: develop another machine that could imitate the changes in the keying procedures that the Germans were making each day.

So a team of Britain's foremost thinkers worked under the most intense security, month after month, in an old Victorian mansion at Bletchley Park, a serene village 40 miles north of London. As time rolled past, the experts began to despair. Then, on the eve of Britain going to war, in September 1939, they hit pay dirt. A device, Ultra, was perfected. It was able to match the electrical circuits of Enigma, permitting Ultra to imitate each change in key procedures by the Germans.[4]

Now, in August 1943, the electronic eavesdroppers in the old mansion at Bletchley Park picked up fragmentary reports that feverish activity had erupted at a locale near Blizna, in southern Poland. MI-6, which had long been in touch with the underground Polish Home Army in Warsaw, asked the Poles for help in reconnoitering the secret German facility.

Disguised as farmers, laborers, and government workers, Polish underground agents, at enormous risk, began snooping about the vicinity. Blizna, it was found, had been cleared of its Polish residents, and German army men and civilians began arriving in droves. Ack-ack guns and searchlights ringed the site, and a railroad spur was being built to connect remote Blizna to the outside world. The Germans took no pains to mask the name they had given to the place—*Artilleria-Zielfeld Blizna* (Blizna Artillery Training Ground).

Early in 1944, Polish underground agents prowling about the Blizna region suddenly heard a loud boom, followed by a screeching noise, and looked up to see a large, strange projectile trailing a fiery plume leap skyward from a clearing in a forest. Word of the electrifying discovery was flashed to London.

MI–6 requested more information. A Polish underground agent (code-named Makary) boldly crawled up to the railroad track leading into the Blizna testing grounds. German guards were but a few yards away. Sitting on the tracks was a German train (its freight cars were marked "dry goods"). Makary noticed in particular an object mounted on an unusually long flatcar. Although covered by tarpaulins, it appeared to Makary to be a monstrous torpedo, one that could sink the largest aircraft carrier or battleship. Within hours, Stewart Menzies had a report on Makary's discovery on his desk at MI–6 in London. This object was no torpedo, Menzies was convinced, but rather a V–2.[5] Further probing by the Polish underground disclosed that the impact area for the missiles was in the Pripet Marshes, 200 miles to the northeast.[6]

Meanwhile, members of the French underground network Century reported on feverish but curious German activities near the hamlet of Mimoyecques, a short distance inland from the port of Calais and 90 miles as the crow flies from London. More than 5,500 engineers, technicians, and excavators had been sent there to build elaborate underground installations for one of the most incredible weapons of the war—the "London Gun."

At first, British intelligence thought that the earthworks were connected to long-range missiles. But other evidence, including that received from the Baltic Sea locale where the London Gun had been test-fired, revealed that these were no doubt installations for those weapons.[7]

Hitler had approved the weapon in July 1943, after Albert Speer, the armaments chief, assured him that the gun would work. Consequently, the London Gun project was given the highest priority in concrete, steel, and manpower. There were to be built and installed along the French coast 50 London Guns, each with a barrel 416 feet long (almost a football field and a half in length). Through the ignition of explosives placed at intervals along the barrel, the weapon would hurl shells for a distance of 100 miles.

Firing tests on one London Gun were held at Hillersleben, a remote locale on the Baltic Sea, on October 19, 1943. They were relatively successful; they managed to fling shells the equivalent of two-thirds of the distance to London.

All the while, the beehive of construction activity at Mimoyecques continued full blast. There, installations for the 50 London Guns were being built underground and protected by a concrete roof 18 feet thick. Only the ends of the muzzles would protrude above ground, and eight-inch thick steel doors would slide across the muzzles when an Allied bomber fleet was approaching.

One hundred feet below the surface was a maze of ammunition dumps, quarters for the crews, and a railroad line that brought in the thousands of tons of shells. And 320 feet below that were the breeches of the London Guns. Each of the monster weapons would have its own elevator and electrical hoists. When all the weapons were firing, they could deluge London with shells at the rate of about 600 per hour.

This enormous installation required enough power for a city of some 60,000 people, so the Germans contacted the Société Électrique du Nord-Quest to supply the power and to erect cross-country power lines. An employee of the Société Électrique du Nord-Quest was also an agent for the Century underground network, and word was flashed to London about the suspicious activities at Mimoyecques.

In November 1943, a U.S. bomber force lifted off from airfields in southern England to pound the earthworks at Mimoyecques. Lieutenant Joseph P. Kennedy, Jr., the eldest son of the ambassador to England in 1940, and the brother of a future president of the United States, was in one four-engine B–24 Liberator loaded with 22,000 pounds of explosives.

Within seconds of Kennedy's takeoff, a second Liberator rose into the sky. The plan was for Joe Kennedy to fly to a point close to the earthworks; then he and his co-pilot were to bail out. That left their aircraft under the radio control of the second bomber which would direct it to crash on the deep installation at Mimoyecques. Not far over the English Channel, Lieutenant Kennedy's airplane (for reasons never officially established) blew up, killing both men on board.[8]

Other bombers, contested every mile of the way by Luftwaffe fighter planes, plastered the target, causing such extensive damage that work on one 25-gun battery had to be abandoned and efforts concentrated on the completion of installations for the second battery.

Adolf Hitler called his London Gun the V-3. Its technical difficulties were never resolved, and a group of nearly 50 scientists who had been working on the project called on the feuhrer in Berlin to break the bad news.

Until December 1943, the various British and American investigations of and operations against suspected German missiles and flying bombs had been largely informal and uncoordinated. Now Hitler's secret weapons threat had become so frightening that the Bodyline committee, chaired since the previous May by Duncan Sandys, was dissolved and a much larger joint British-American group code-named Crossbow was set up in its place. Crossbow would encompass all phases of operations against the German long-range weapons program—research, development, experimentation, production, construction of launching sites, transportation, and tactical firing.

Urgency was injected into Crossbow on New Year's Day, 1944, when Joseph Goebbels took to Radio Berlin to launch a tirade against the Western Allies and to trumpet to the Herrenvolk (people) that the fuehrer's secret weapons soon would convert London into a pile of ashes.

All the while, operatives in MI–6, perhaps the world's most effective secret service agency, were laboring arduously to hatch a plot that would unlock the mystery of the fuehrer's vengeance weapons. They mulled over at length a plan to kidnap a German rocket scientist but abandoned that angle when it was learned that the human target would be closely guarded by SS troops.

# 6

# The Gestapo Strikes at Midnight

Reichsfuehrer Heinrich Himmler, a one-time chicken farmer, was the most feared man in Germany and its second most powerful figure. As minister of the interior, commander of the elite SS, chief of the Reserve Army, and head of the Gestapo and all German police forces, Himmler wielded gargantuan clout. It was known throughout the Wehrmacht and government that Himmler had compiled thick dossiers on each leader and, when the time was ripe, would use the damaging materials (much of it rumor and gossip) to frame a rival. One of those Himmler had targeted was Wernher von Braun, who was unaware that he incurred the reichsfuehrer's wrath.

On February 21, 1944, von Braun had just landed his airplane in the darkness of Peenemünde after a visit to the Blizna testing site when he was handed a telephone message that had arrived two hours earlier. He was ordered to report to Reichsfuehrer Himmler immediately. Von Braun was puzzled: he worked for the army, not for the SS.

Nonetheless, no one could ignore an order from Himmler. Shortly after dawn, von Braun wearily climbed back into his Messerschmitt–108 Typhoon and flew alone to Hochwald, East Prussia, where Himmler's field headquarters was located in his personal luxury train. When the young rocketeer entered Himmler's office, he was greeted with icy politeness. The reichsfuehrer, von Braun reflected, was as mild-mannered a villain as had ever slit a throat and resembled a stereotype schoolteacher. Himmler, his steely eyes peering through his thick pince-nez which gave him an owlish appearance, wasted no time on idle talk.

"I hope you realize that your [V–2] rocket has ceased to be a toy," the reichsfuehrer said. "And the whole German people eagerly await the mystery weapon. . . . As for you, I imagine that you've been immensely handicapped by army red tape." Then Himmler dropped his blockbuster: "Why don't you transfer from the army to the SS? I have access to the Fuehrer, and I promise you vastly more effective support than can those hidebound army generals."[1]

Taken aback, von Braun replied, "Herr Reichsfuehrer, I couldn't ask for a better chief than General Dornberger. Such delays as we're still experiencing are due to technical troubles and not red tape. You know, the V–2 is rather like a little flower. In order to flourish, it needs sunshine and a gentle gardener. What I fear you're planning is a big jet of liquid manure! You know, that might kill our little flower."[2]

Himmler's face flushed and he smiled a bit. It was a facial expression that many high officials had learned to fear.

Piloting his Messerschmitt back to Peenemünde, von Braun felt uneasy about his encounter with Himmler. But he plunged back into his V–2 research and the encounter at Hochwald soon slipped from his mind.

A month later, on a Sunday in early March 1944, von Braun felt the need for a brief respite from his endless grind of 15-hour days and attended a cocktail party held at a private residence at a seaside port on Usedom island. Also present were several scientists and engineers from Peenemünde, Wehrmacht officers, and civilian guests.

Von Braun, a witty and jovial man, played the piano (on which he was quite gifted), then joined two fellow rocketeers, Klaus Riedel and Helmut Grötrupp, in a corner of the room. High-spiritedly and naively, they indulged in their dreams and talked excitedly about how the V–2 could be developed ultimately for space travel—perhaps even reaching the moon one day. Von Braun conjectured that a rocket might be developed to carry mail between Europe and the United States after the war.

Discussions about space travel were not unusual among the Peenemünde rocketeers; discussions had gone on for years, even though von Braun had been cautioned many times by Walter Dornberger to soft-peddle such talk. "You never know who might be listening," Dornberger had declared.

Von Braun ignored the advice from his older friend and boss. In fact, the young scientist even had permitted a German picture magazine to prepare a layout with all sorts of fantastic drawings based on his visions of eventual space travel and explorations. One especially outrageous illustration showed two men actually landing on the moon. Presumably this magazine layout found its way into the Wernher von Braun dossier in Reichsfuehrer Himmler's headquarters at Prinz Albrecht Strasse 8 in Berlin.

It was just past midnight on March 15, 1944, a few days after the convivial party on Usedom island, when von Braun was awakened at his bachelor quarters by a sharp and persistent banging on his door. Sleepily opening the door, he saw three stern-faced men in civilian clothes and immediately sensed their identity—the Gestapo.

"*Anziehen Sie sich and kommen Sie mit uns!*" (You will dress and come with us!) one man exclaimed. Von Braun was stunned. "But why?" he asked. "I am needed here at my work." "We have our orders," the Gestapo agent said. "You will come with us at once to the *Polizei Prasidium* [police station] in Stettin."[3]

Accompanied by the three Gestapo men, von Braun was driven to Stettin, about 70 miles to the south. At the police station, he was in for another shock: his cronies, Helmut Grötrupp and Klaus Riedel already had been brought in.

Von Braun, Riedel, and Grötrupp were put in separate cells where they would remain for two weeks without even a hint about why they had been arrested. Not knowing what their fate would be was an especially gut-wrenching experience for the rocketeers.

Before dawn on March 15 (a day after the arrest of the three rocket scientists), Walter Dornberger (now a major general) was awakened by the impatient jangling of a telephone on his bedside table in his quarters in Schwedt, 40 miles south of Peenemünde. General Walter Buhle, chief of staff of the Oberkommando der Wehrmacht (OKW), ordered Dornberger to report as rapidly as possible to Field Marshal Wilhelm Keitel at Berchtesgaden in the Bavarian Alps of southern Germany. There, perched on the top of a mountain with a breathtaking view, was *Adlerhorst* (Eagle's Nest), Adolf Hitler's retreat.

Buhle refused to disclose the reason for the urgent summons to a locale 400 miles away. Dornberger leaped out of bed, dressed, gathered a few belongings, and scrambled behind the wheel of his car. Departing at 8:00 A.M. on this bitterly cold, blustery day, he drove over icy roads and through snow squalls and reached Berchtesgaden after dark.

General Buhle guardedly told Dornberger that the Gestapo had arrested three of Germany's most prominent rocket scientists, including Dornberger's longtime friend and protégé, Wernher von Braun, and charged them with sabotaging the V–2 program. Stunned by the shocking news, he demanded that Buhle provide him with details, but the OKW officer begged off, saying that Field Marshal Keitel would give him specific reasons for the arrests.

As chief of the Oberkommando der Wehrmacht and Hitler's most trusted military advisor, Keitel had been promoted to field marshal back in 1940 and had been at the fuehrer's elbow—and ear—ever since. Keitel was held in low regard by most German generals and was considered to be a flagrant opportunist, a toady for the gigantic ego of the Reich's top leader. Keitel was said to have once had as his lifetime ambition to become a farmer, and many Wehrmacht generals often expressed the opinion that he had missed his calling. Summed up one general: "Keitel is a blockhead!"[4]

Now, Walter Dornberger was in for another jolt. "The charges are so serious against [the three scientists] that they are liable to lose their lives," Keitel declared. When Dornberger replied that he would be willing to vouch for their total loyalty with his own life, the stone-faced field marshal said in a grave tone: "Do you know that your closest colleagues have stated in company at [the Usedom island party] that it had never been their intention to make a weapon out of your rocket? That they had worked, under pressure from yourself, at the whole business of development only in order to obtain money for their experiments and the confirmation of their theories? That their object all along has been space travel?[5]

Dornberger asked who had made the charges. Keitel replied that he did not know. When Dornberger persisted, the field marshal shrugged his shoulders and said testily: "Look, there's nothing I can do about it. Himmler has taken over himself."[6]

There were a few minutes of awkward silence. Then Keitel, who had risen to his high post and clung to it by the expedient of avoiding clashes with the SS and not confronting Adolf Hitler with thorny problems, suggested that Dornberger be "reasonable."

"I must avoid the suspicion of being less zealous than the secret police and Himmler in these things," the field marshal declared. "You know my position here. I am watched [by the Gestapo]. All my actions are noted. They are only waiting for me to make a mistake."[7]

Then General Dornberger put forth a course of action that few men in Nazi Germany would follow: he asked Keitel to arrange a personal interview for him with Reichsfuehrer Himmler. Keitel put through the call, but Himmler's adjutant replied that the reichsfuehrer would not see von Braun's boss.

Seething with anger, Dornberger hopped in his car and drove through the frigid weather to Berlin, where he called on SS Oberstgruppenfuehrer Ernst Kaltenbrunner, the Reich's chief of security, who referred him to his number-two man, SS Obergruppenfuehrer Heinrich Müller, who was in direct charge of the Gestapo.

Müller pointed out that the arrests had been ordered by Himmler personally and that there was "ample evidence" to convict the scientists. The Gestapo boss, like Himmler, cold, calculating, and ruthless, informed Dornberger that "we have a fat file of evidence against you, too." Müller added, "You, too, Herr General, are under constant surveillance."

Heartsick and frustrated, Dornberger drove back to Schwedt. Anger swelled in him. Germany's armed forces were reeling on all fronts, the homeland was being pounded to rubble by huge Allied bomber fleets, an enormous Allied army was gathering in England to cross the Channel and invade France, yet the Reich's top leaders wasted much of their time and energies squabbling with one another and lusting for ever more power.

While Walter Dornberger was futilely pleading with Kaltenbrunner and Müller, Armaments Minister Albert Speer learned of the arrests while confined to a hospital in Kresshiem with a serious pulmonary ailment. He could not believe his ears. Wernher von Braun had toiled indefatigably, night and day, for ten years to develop the revolutionary missile. Now Himmler was charging him with sabotaging the program.

By happenstance, Adolf Hitler found time from his backbreaking schedule to pay an unexpected visit to Speer's bedside. Speer seized this golden opportunity to plead for the release of von Braun and the other jailed scientists, reminding the fuehrer that von Braun was the genius behind the V–2 and that the entire missile program could be jeopardized without the young scientist being available to smooth out the final kinks in the weapon. Von Braun's absence, Speer stressed, would be a serious blow to the Third Reich's war effort.[8]

Two weeks after his arrest, the guards at the Stettin police station hauled Wernher von Braun from his frigid cell and took him to a small room where SS officers were seated behind a simple wooden table. Himmler's men were to be the judges in a trial to weigh charges brought against von Braun by Himmler, their boss.

For the first time the accused learned why he had been arrested, for sabotaging the war effort by declaring that he had never intended to develop the V–2 as a weapon but rather had intended to use it for space travel. As von Braun suspected, Himmler or his minions had planted among the guests an informer who had eavesdropped on the casual conversation among him, Klaus Riedel, and Helmut Grötrupp at the cocktail party a few weeks earlier. That informer, it was later disclosed, was a woman dentist.

A second charge was lodged against von Braun—treason. The "court" maintained that von Braun kept an airplane in readiness at all times in order to flee to England and hand over top-secret materials from his rocket research to British intelligence. Von Braun could have had a difficult time trying to prove that he was not thinking about defecting to the enemy, for he usually kept a small government plane which he piloted himself on business trips all over Germany. It would be a simple matter to lift-off and simply steer a course to England. How could he prove that he had no traitorous intentions?

In addition to sabotaging the war effort, Klaus Riedel and Helmut Grötrupp also were accused of being pacifists.

Pale-faced and racked with tension, the three defendants awaited their fate. Each reflected whether a guilty verdict would mean a firing squad or a hanging. Just then, in dramatic fashion, General Walter Dornberger burst through the door of the "courtroom," strode briskly up to the presiding SS officer, and handed him an official document. The SS man swallowed hard: the signature read "*A. Hitler.*"

All charges were immediately dismissed. Albert Speer's intervention from his hospital sickbed had saved the day—and the three rocketeers' necks—just in the nick of time.

Von Braun and his two associates returned to Peenemünde to resume their research, but their arrests had created a climate of fear among civilian technicians that seriously delayed the refinement of the V–2, on which Hitler had placed his hopes for ultimate victory. From a personal viewpoint, von Braun came away from Stettin with a deep aversion to police states—and to Nazis (although he himself had been an inactive member of the party since his early 20s).

Meanwhile, the Harz Mountains of central Germany had been transformed into a beehive of activity in the wake of the fuehrer's decision that V–2 missiles would be mass-produced there in what would become the largest underground factory in the world. Back in 1933, the huge conglomerate, I. G. Farbenindustrie, had begun drilling into the rock and installing fuel tanks there. Over the years, the drilling had continued.

In September 1943, General Hans Kammler, chief of the building branch of the SS's main office, was put in charge of all construction work for the missile program as the result of the new powers bestowed upon Reischsfuehrer Himmler by the fuehrer following the RAF bombing of Peenemünde. Up to this point, construction for the missile program had been an army responsibility. Himmler had edged his foot a little farther into the V–2 door.

Nearby, a camp (code-named Dora) was established, and into it poured a steady stream of political prisoners—Poles, Russians, French, Italians, Belgians, Dutch, Czechs, and Germans. By January 1944, there were 11,000 prisoners in Dora. Hundreds of them were used to enlarge the underground factory, and it was transformed into a maze of 46 tunnels, each 250 yards long, 14 yards wide, and up to 30 yards high, bisected by a pair of tunnels two miles deep. Ventilation shafts changed the air each day, and special lighting kept the entire underground cavity illuminated. Most important, the factory was safe from Allied bombing.[9]

This mammoth underground factory belonged to *Mittelwerk*, a company established for the mass production of the V–2 and the flying bomb. Its main office was in Berlin. The Luftwaffe had perfected the flying bomb, and it was rolling off assembly lines at the Volkswagen factory in Fallersleben.

Although the V–2 still had defects that often caused it to explode on launching or in the air a short distance before reaching a target, haste was the password and mass production began. On January 2, 1944, the underground factory shipped out the first three long-range missiles.

At Berchtesgaden, Adolf Hitler, a supreme optimist, still retained absolute confidence that his Wehrmacht in the West would smash the looming Allied cross-Channel smash into France. His critics—Allied as well as German—sneered at what they called "Hitler's intuition." Yet the fuehrer, whose only military service was as a World War I infantry corporal, was a shrewder tactician than his detractors would admit.

On May 16, 1944, the fuehrer was certain that waves of his flying bombs would turn the tide in favor of the beleaguered Third Reich, so he issued orders concerning the bombardment of London. The precise time for launching the robot offensive would be given by 67-year-old Field Marshal Gerd von Rundstedt, *Oberbefehlshaber Westen* (commander in chief, West), whose headquarters was in a Paris suburb.

Von Rundstedt, known in the Wehrmacht as "the last of the Prussians," would coordinate the flying-bomb offensive. Heavy Luftwaffe air attacks and a bombardment of British southern coastal towns by long-range conventional artillery in the Pas de Calais would be launched at the same time as the robots.

Hitler knew that the 692 square miles of Greater London would be a target difficult to miss, despite the relative inaccuracy of the flying bombs. The robot onslaught, the fuehrer was convinced, would shatter the Allied invasion before it began. However, his intuition (and intelligence reports) failed him, for he set the tentative date for beginning the flying-bomb assault on June 16, ten days after the Allied high command had set D-Day.

Three minutes after midnight on June 6, 1944, hundreds of Allied transport planes began disgorging thousands of paratroopers behind German defensive positions along the Channel coast of Normandy. The long-awaited invasion of Hitler's *Festung Europa* (Fortress Europe) had been launched. Amphibious assaults hit the beaches just after dawn, and by nightfall, 72,135 American and 82,115 British and Canadian troops had carved out a toehold in France.

# 7

# The Polish Underground Steals a V–2

Just before 9:00 P.M. on June 12, 1944—D-Day plus 6 for the Allies—Colonel Max Wachtel was handed an urgent message from Field Marshal Gerd von Rundstedt that had just arrived on the *Blitzfernschreiber* (high-speed teleprinter). Wachtel, a tall, robust man of 45, was commander of Flak Regiment 155 (W), the code name for the flying-bomb batteries arrayed along the Channel coast of France.

Von Rundstedt's signal contained one phrase: Junk Room. The robot onslaught against London was to begin—immediately. Much to the delight of Reichsmarschall Hermann Göring and the Luftwaffe, it would be their flying bomb, not the army's V–2 missile, with which Adolf Hitler would strike his first retaliatory blow.

For the fuehrer, there was no time to lose. About one million Allied soldiers were ashore in Normandy and threatening to break out of their beachhead and dash for Paris and then on to Berlin.

Wachtel's superiors, feeling the hot breath of Adolf Hitler on their necks, had been hounding the Flak Regiment 155 (W) commander daily. They accused him of being slow and careless and threatened to court-martial him for sabotaging the war effort. None of the generals over Wachtel understood his tribulations, and when he tried to explain them, they cut him off abruptly. So effective had been the Allied air forces' pounding of the launch sites, supply trains, and storage depots that the Germans had only ten robots ready for launch when von Rundstedt flashed the signal Junk Room.

Across the Channel, British intelligence expected a deluge of flying bombs and the government had been secretly preparing its defenses. Not only had MI–6 collected much physical, wireless, and technical evidence that the robot assault was impending, but the Germans themselves unwittingly had tipped off the Brits that the onslaught was at hand.

Nearly five years earlier, within hours of England's declaration of war against the Third Reich in early September 1939, agents of MI–5 (Britain's counterintelligence agency) and Scotland Yard began fanning out over the land in a mammoth roundup of German spies. The spybusters had their hands full: there were 356 names on their Class A espionage list, plus hundreds of others suspected of spying for Hitler. Actually, the Abwehr had 256 spies in Great Britain, many of whom had been deep undercover for years. There were at least ten women in the Abwehr spy apparatus in Britain, including two women in their 50s who worked as maids in the homes of two British admirals.

As the weeks and the months went by, MI–5 and Scotland Yard rounded up nearly every Abwehr agent in England. When key spies were collared, MI–5 confronted them with a choice: they could be hanged or they could become double-agents and send back misleading information to their controllers in Germany. Without exception, each spy chose the latter course and continued to dispatch intelligence reports cagily orchestrated by MI–5 on the AFu radios the Abwehr had provided them.

Unaware that their star spies had been captured and "turned" long ago, the Abwehr, in early June 1944, radioed for them to hightail it out of London as rapidly as possible. This was the final clue British intelligence needed to conclude that a buzz-bomb attack was impending.

Seven hours after Colonel Wachtel received the Junk Room order, an elderly volunteer of the Royal Observer Corps in Kent, England, was peering skyward; it was 4:00 A.M. He was startled to see a strange object streaking through the dark sky toward London. Its dim contours resembled those of a miniature airplane, but its exhaust was belching reddish-orange flame, and it gave off a sputtering sound like that of an old automobile chugging up a steep hill.

A few minutes later, the robot's preset timer cut off the engine, and the V–1 plummeted onto the sleeping village of Swanscombe, some 18 miles from its target, the Tower Bridge in London. An enormous explosion rocked the region. Within the hour, three more robots hit, at Cuckfield, Platt, and Bethnal Green.

During the night, ten robots screeched off launch ramps in the Pas de Calais region of eastern Normandy. Of these, four crashed on launch, two plunged into the Channel, and four got to Greater London. The initial attack was hardly the hurricane of 1,000 V–1s that the fuehrer had demanded.

When Winston Churchill convened his War Cabinet later that day, most of those present were jubilant. Hitler's ballyhooed secret weapon was a monumental fizzle. Casualties had been small (six Brits killed), the robots' precision had been poor, and instead of the 400 flying bombs in the initial assault predicted by British intelligence, only four had hit. Many at the meeting felt that if the secret weapons assault against Britain, about which Joseph Goebbels had been crowing for months, was such a failure, then the vaunted V–2 missile threat also must be merely a propaganda device. Alexander Cadogan, the undersecretary of the Foreign Office, gleefully scrawled in his diary: "Not very impressive. Hope some will return and fall on Germany!"

Young Reginald Jones of MI–6 did not share in the optimism and felt that the opening salvo was a misfire that the Germans would correct. He telephoned Lord Cherwell, Churchill's scientific advisor, and urged the War Cabinet to proceed with caution. Cherwell, too, was exuberant over the robot-bomb fizzle, no doubt feeling that his often voiced view that Hitler's secret long-range missiles and flying bombs were merely bluffs had been vindicated.

Cherwell "chuckled exuberantly" and told Jones that "the mountain hath groaned and given forth a mouse!" Jones replied: "For God's sake, Lord Cherwell, don't laugh this one off!" Cherwell chuckled again.[1]

For three days, not another buzz bomb was launched. Then, at noon on June 15, and continuing relentlessly for 24 hours, 244 V–1s flew off their ramps, and 144 of them reached England. Seventy-seven of the robots dived on Target 42, the German code name for London.

When Winston Churchill and the War Cabinet met on the following morning, the jubilation had vanished. Replacing it was deep worry. General Eisenhower, whose armies were still struggling to stay ashore in Normandy, was asked to take "all possible measures to neutralize the supply and V–1 launching sites [along the French coast] subject to no interference with the essential requirements of the Battle of France."[2]

In this period of deep anxiety in official London, the Polish Ministry of Defense in the British capital radioed an urgent coded signal to two of its crack spies in northern France, Lieutenant Wladyslaw Wazny (code name *Tygrys*) and Sergeant Edward Bomba (code name *Toreador*), who were on the run from the Gestapo. The two Poles were asked to direct immediately all of their attention to locating the launch ramps of the buzz bombs. They were also to report on the bomb depots and the means of transportation. Finally, they were to send back detailed reports on the amount of destruction or damage inflicted upon the ramps and depots by impending Allied bomber attacks.

Wazny and Bomba had parachuted into France the previous March and were followed by four Polish radio operators in several night drops from British aircraft. Now, as ordered, the Polish network began scouring the countryside along the French coast. From a Polish national who had been conscripted into the Wehrmacht, Wazny and Bomba picked up a crucial point: any large yellow road signs with black trimming meant that a flying-bomb launching pad was nearby.

When details of a launching site were sniffed out by Wazny and Bomba, a report was coded and sent by courier to the radio operators, who had to keep on the move constantly in order to avoid the Abwehr's electronically equipped vans that crisscrossed the region to snoop out clandestine radio transmitters.

In the weeks ahead, the Polish underground network radioed London reports on the precise locations of more than 150 launching pads (known to the Allies as "ski jumps" because of their contours), as well as two-score reports on depots and V–1 transportation facilities.[3] Despite massive Allied bombing of the ski jumps along the French coast, Flak Regiment 155 (W) launched more than 2,000 robots against England by June 25.

British civilians lived in constant fear of the grating sound that resembled a badly tuned motorcycle, although there was little panic. Tens of thousands were killed or wounded, and perhaps 200,000 houses and other structures were destroyed or badly damaged. The threat of an epidemic of illnesses due to smashed water lines and sewers increased official worries.

An evacuation program for London children, some women, the elderly, and the sick was rapidly organized. The mass exodus from Target 42 into the countryside resulted in a severe strain on British railways and hampered the vital flow of ammunition, weapons, and supplies to the fighting men in Normandy.

Adolf Hitler, in his thirst for vengeance for the Allies' 1,000-bomber raids on Berlin, had made a monumental mistake in directing the buzz bombs on London. Had the robots been launched against the southern Channel ports—Plymouth, Southampton, Dover, Margate, Portsmouth, and Torquay—the damage and casualties might well have drastically disrupted the Allies' supply schedules in support of the troops in Normandy and permitted German Field Marshal Erwin Rommel to regain the initiative.

In the meantime, Prime Minister Churchill proposed to the War Cabinet that Britain retaliate for the indiscriminate robot assault against civilians by the all-out use of poison gas, mainly mustard gas, on the Germans. The suggestion was dropped, however, when the British military high command responded that "the use of [poison] gas in Europe would achieve an initial tactical surprise, but would thereafter restrict Allied tactical movements."

Launching bacteriological warfare against the Third Reich also was considered seriously by the War Cabinet at a tense meeting on July 16. Earlier, British scientists had developed a bacteriological agent code-named "N," for which there was no known remedy or antidote. "If 'N' is used in practice," a report said, "the effect on [German] morale will be profound. . . . [I]t might lead to a breakdown in [enemy] administration with a consequent decisive influence on the outcome of the war."[4] Retaliating for the robot attacks by initiating bacteriological warfare was ruled out, however, because "no sustained attack with 'N' [was] possible before the middle of 1945."

Churchill also concocted another retaliatory scheme. He would select 100 small, undefended German towns, and Royal Air Force bombers would wipe them out, one by one, "until Herr Hitler calls off his dogs." That plan also was rejected.

In the meantime, a lone MI–6 agent in northern France, a British lawyer, W. J. Savy (code name Wizard) pulled off an intelligence coup that would have been rejected by Hollywood scriptwriters as implausible. His deed vastly diminished the robot assault on London.

After parachuting into France prior to D-Day, Wizard went to ground in Paris, and on a trip to nearby Creil, he engaged in casual conversation with a Frenchman, who said he grew mushrooms in huge caves at Saint Leu d'Esserent, a short distance away. Lawyer Savy, employing his slickest courtroom manner, drew out the mushroom man and learned that the Germans had taken over the caves and had built a railroad line leading up to them.

Wizard's ears pricked up. Why would a railroad have to be constructed, in wartime, for the purpose of moving mushrooms? What's more, the Frenchman said, the roofs of the caves had been lined with concrete and braced with steel pillars. This had to be some kind of an arms dump, Wizard reflected.

At great risk to his health, Wizard stole a look at the premises leading up to the caves and, along with cautious inquiries among French civilians in the region, discovered that the Germans were storing in the caves perhaps 2,000 flying bombs. This electrifying information was radioed to London by Wizard.

On the night of July 4, 227 British Lancasters dropped 4,000-pound block-busters on the caves and the rail line leading up to them. Luftwaffe fighters promptly rose to the attack and, swarming like angry bees, shot down 13 bombers. However, another flight of Lancasters, each carrying a six-ton "earthquake bomb," followed and plastered the caves with 11 earthquakes, caving in the huge cavities.

Wizard's mushroom man, a day later, reported that the entrance to the caves had been blocked but that the Germans already had commenced opening them and digging out the many robots that were buried inside by earth. During the next 24 hours, the Germans finished the arduous task.

Wizard flashed word to London, and on the night of July 7–8, some 330 RAF heavies paid a return call to the caves. This time the air armada winged in extremely low and deluged the entrances and nearby V–1 dumps in a flaming mass of high explosives. This time, the job had been completed, and it denied to Colonel Wachtel nearly 25 percent of the buzz bombs available to him in France.

Meanwhile in Great Britain, the supersecret XX-Committee (Double-cross Committee) was racking its brains for a scheme to confound the Germans over how accurate their buzz bombs were and to cause the weapons to impact elsewhere than in teeming London. Back in 1940, Winston Churchill had conceived the XX-Committee, within MI–5; its function was to develop and implement diabolical deception hoaxes to confuse Adolf Hitler and the Abwehr.

Unaware that its spies in London had years earlier been captured and "turned" by British counterintelligence, the Abwehr now urgently radioed their agents to hurry back to the British capital (they had never left it) and send Berlin reports on where the flying bombs were landing. This unexpected turn of events gave the XX-Committee the opportunity it had been seeking. But the XX-Committee was confronted by a dilemma. On the one hand, if the double-agents were allowed to radio accurate reports on the buzz bomb landings to their former masters in Berlin, they would be providing enormous aid to the enemy. Colonel Wachtel and his gunners would continue to aim at the Tower Bridge with the assurance that the robots were falling somewhere in London. On the other hand, if the turned spies lied about the impact spots, Luftwaffe photoreconnaissance would disclose that they had done so, and the Abwehr finally would suspect that large numbers of their spies had been turned and were feeding the Third Reich faulty—and damaging—information contrived by British counterintelligence.

What was more, Colonel Wachtel's officers in Flak Regiment 155 (W) kept a record of the precise second that each buzz bomb was fired. While they could not be certain where the explosion had occurred, they knew the time of impact within one or two minutes.

Reginald Jones now proposed an ingenious deception hoax. The double-agents would report to their masters in Berlin the site of robots that had landed north of London, using the impact times of missiles that actually had hit south of the capital. Jones hoped that the Germans would conclude that the buzz bombs were landing long and shorten the range by adjusting the engine cut-off devices to decrease the flight times. That, conceivably, would result in the robots hitting in open fields south of London.

Reginald Jones' plan went up the pecking order to the War Cabinet, which approved it, and the XX agents began radioing back the phony reports concocted by their British controllers.

Jones' plan was not an unbridled success. As anticipated, the Germans did shorten the buzz bombs' range, but not all of the robots hit in open country; some hit the blue-collar residential areas south of the Thames River. Home Secretary Herbert Morrison protested vigorously, but his pleas were ignored and the Jones deception scheme remained in force.

Although the Germans had shifted long-range missile firing tests to Blizna in Poland, the urgencies of war dictated that experimental launchings also be held at partially rebuilt Peenemünde East on a much smaller scale. On June 13, a Wasserfall antiaircraft rocket equipped with a revolutionary radio mechanism was prepared for firing. The mechanism allowed the Wasserfall rocket to be steered by remote control, but the device was not designed for use on the V–2.

This Wasserfall soared high into the blue heavens, but it went awry; instead of splashing into the Baltic Sea, it flew too long and landed near Malmö, in southern Sweden. Authorities there were quickly notified and the Swedish air force put an armed guard around the impact site and the rocket.

Nearly a month later, on July 10, Stewart Menzies, chief of MI–6 in London, received a report that the Wasserfall had fallen in Sweden and that the Germans were desperately trying to recover it. One group of Germans tried to enter the sealed-off area behind a slowly moving hearse, posing as mourners. Swedish guards turned them away.

Menzies promptly entered clandestine negotiations with Swedish officials. Sweden was a neutral country in the war, but it was more neutral toward the Western Allies. Finally, a deal was struck. In return for the missile's remains, Britain would ship two squadrons of tanks to the Swedish army. Two unarmed C–47 transports carrying ordnance officers flew to Sweden, where the Wasserfall was dismantled, put on board, and taken back to England.

Finally, the British scientists had received what they had long desperately sought—a complete German missile. At the Royal Aircraft Establishment (a government research facility) at Farnborough, outside London, a huge tent was erected and the hundreds of pieces of the Wasserfall were spread out on the ground.

The radio apparatus, adapted to remote control, was reconstructed, and Reginald Jones and other British scientists reached the conclusion—a false one—that the V–2 missiles would be dangerously accurate for they also would be equipped with this same device and could be guided to their targets. Unknown to the Brits at the time, there was a radio apparatus on the V–2s, but it was designed only for receiving and transmitting signals, not for remote control.[5]

Based on their findings with the Wasserfall, Reginald Jones and other experts at Farnborough concluded that it might be possible to develop a radio beam to deflect off-course an approaching V–2. Consequently, they began prodding Stewart Menzies at MI–6 to provide them with a V–2 for analysis. It was a tall order. Even with its traditional reputation for ingenuity, MI–6 could not conceive of a means of stealing a 46-foot, 13-ton missile from under the noses of vigilant German security people and transporting it to England.

MI–6 sent an urgent request to the secret Polish underground headquarters in Warsaw: could the Poles send an intact V–2, or even parts of one? It was an outrageous request. Blizna was so tightly guarded by SS troops that the British might just as well have asked the Poles to smuggle out Adolf Hitler's desk at Wolf's Lair.

However, within days a fortuitous event occurred—a V–2 that had gone awry after being launched at Blizna plunged into the swampy bank of Poland's Bug River and failed to explode. Its nose was buried deep in the mire, but the missile was intact. The rocket was spotted by a farmer who lived nearby, and he excitedly reported it to a member of the Polish underground.

News of the electrifying discovery was flashed to the Polish Home Army headquarters in Warsaw. A young engineer, Antoni Kocjan, promptly took two men and rushed to the Bug River to photograph the V–2. It was risky business, for German patrols already were literally beating the bushes in a frantic effort to locate the errant missile. Before leaving, the partisans covered the projectile with brush and foliage.[6]

A few nights later, Antoni Kocjan and a few Polish scientists stole through the blackness to the Bug River on a strange mission: to steal key parts of a 46-foot missile. Constantly in fear of detection by the Germans, the Poles worked feverishly through most of the night and, finally, with the aid of three teams of husky, snorting plow horses, pulled their coveted prize out of the sticky morass of mud. The engine and steering mechanisms were removed by the scientists, piled onto farm carts, trundled across the rough fields, and hidden in a nearby barn.

After daybreak, the scientists took many photographs and measurements of components, made drawings, and then dismantled the parts into hundreds of pieces, large and small. These components were lifted onto two ancient trucks and covered by loads of potatoes.

Driving toward Warsaw in the wheezing, coughing trucks was a frightening experience. The vehicles were halted at three different roadblocks. While the resistance men in the cab sat motionless, their hearts thumping furiously, the enemy soldiers studied their potato cargoes, poked bayonet-tipped rifles into the vegetables several times, then permitted the trucks to continue.[7]

In London, Polish intelligence was elated over the bonanza dragged out of the Bug River and set into motion a plan to smuggle the key components out of Warsaw and back to England. Wildhorn III was the code name for the perilous operation.

Just past 10:00 P.M. on July 25, a lumbering C–47 transport plane (a U.S. product known to the British as a Dakota), lifted off from Brindisi airfield in Italy and set a course for an abandoned airfield (code-named Moytl) in Poland. Operation Wildhorn III (the recovery of the V–2 components) was underway.

The pilot was Flight Lieutenant Stanley G. Culliford, a New Zealander, and the co-pilot was a Pole, Flight Lieutenant Kazimierz Szrajer. The two officers and their British crew held no illusions that their secret mission would be a "milk run," as airmen labeled a routine flight. The C–47 was to land directly in the center of territory crowded with German units retreating westward under pressure by the advancing Russian army.

It was just past midnight at the dark airstrip, once used by the Polish air force, when a waiting band of armed partisans heard the faint engine hum of an approaching airplane. The men were tense. That day, German army units had moved into villages in the area; one enemy force was bivouacked only a half-mile away. Off in the distance could be heard the muffled boom of artillery, for the fighting front was not far away.

Now the Dakota circled overhead and the partisans (who had been notified by code phrases over BBC radio that it was coming) recognized the dim silhouette as that of a two-engine transport plane. Lieutenant Culliford began gliding in to a landing when suddenly the dark locale burst into brightness—the pilot had turned on his headlights. The partisans cringed: if the roar of the engines did not alert the Germans, no doubt this glare of iridescence would do so.

The plane rolled to a stop and shadowy figures stole silently from surrounding woods. They were pulling carts loaded with several large containers that had served as oxygen drums. In the drums were the key V–2 components. The precious cargo was loaded into the plane, and four Polish scientists from Warsaw climbed aboard. Culliford revved the engine for takeoff—but the C–47 refused to budge. Something had gone wrong. Tension gripped everyone present.

Lieutenant Culliford dashed back from the cockpit and called out in a stage whisper for everyone to get out. The crew hastily inspected the wheels and found that they were bogged down in mud. Frantically, dirt was shoveled away in an effort to free the wheels. Then everyone scrambled back into the airplane once more. Again the raucous roar of revving engines. Again the plane refused to move. Everyone out, Culliford ordered.

Now the tension was nearly unbearable. Passengers and crew jumped onto the landing strip. It was found that the wheels had sunk even deeper into the mire. Visions of Gestapo torture chambers danced before their eyes—at any moment the Germans, attracted by the turmoil, might race up.

Gather up sticks, was the order. Dark figures dashed about helter-skelter. The sticks were lodged under the wheels. Everyone back on board. It was not easy getting in and out of the aircraft in the blackness, for each passenger was burdened with suitcases and boxes holding intelligence documents and other materials stolen from the Germans and others relating to the V–2. Once more the harsh revving of the engines. This time, the noise seemed louder than ever. The C–47 shuddered violently, seemed ready to bolt forward—and remained stuck. Everyone out!

In desperation, Lieutenant Culliford switched on the headlights to allow crewmen to see the wheels more clearly. This brought a renewed surge of cold fear into hearts. The wheels had sunk yet deeper into the mud, and Culliford gave the order to blow up the airplane.[8]

With heavy hearts the partisans and scientists watched silently. All that painstaking work. All that peril. Had it been in vain now that they were so close to success? As the crew placed the explosives, a few partisans began digging frantically around the wheels with their bare hands. They clawed and clawed some more. Blood trickled down their palms. Other partisans ran to a fence and brought back several wooden slats which were wedged under the wheels. Now the explosives were ready to be ignited. The Poles begged Culliford to try to take off one more time. Everyone scrambled back into the C–47.

Now, the abused engines grew obstinate. They coughed and sputtered and groaned. Then the engines turned over and roared loudly.

Suddenly, the aircraft pulled loose from its vise of sticky mud, rolled down the runway, and lifted into the air. A flood of cheers from passengers and crewmen rocked the plane. Glancing back, Lieutenant Culliford saw a string of vehicle headlights moving toward the airstrip—German soldiers, no doubt.

Forty-eight hours later, the C–47 landed in England, and British scientists began studying the V–2 components. Earlier they had theorized from available intelligence that the missiles' flight course could be altered by radio waves, but they soon learned that this would be impossible, because the V–2 mechanism did not react to countermeasures by radio. There would be no defense against a supersonic missile onslaught.

Great Britain was in mortal peril.

# 8

# The Great Treasure Hunt Begins

British intelligence was unaware that Reischsfuehrer Heinrich Himmler was now in charge of the V–2 program—lock, stock, and swastika. After 36-year-old Lieutenant Colonel Klaus Philip Maria Count von Stauffenberg, a key member of the conspiratorial group Schwarze Kapelle, narrowly failed to blow Adolf Hitler to pieces with a time bomb placed under a conference table at Wolf's Lair on July 20, 1944, the fuehrer no longer trusted his army generals.

As soon as Himmler had been bestowed the new power, he promoted Hans Kammler, his protégé, from SS brigadier general to SS major general and gave him direct control of the entire missile project, from development to firing. Kammler's new title was "special commissioner." "My orders and [Kammler's] directions are to be obeyed without question," Himmler's order declared.

General Walter Dornberger, who had given birth to the army's rocket program back in 1930 and nurtured it to the brink of phenomenal success, was given the menial title "technical staff officer." General Kammler was now Dornberger's boss.

Dornberger knew that his dealings with Kammler would be rancorous ones. Only a few weeks earlier, the young SS general harshly snapped at Dornberger that he should be court-martialed for squandering so much of the Reich's money and manpower trying to make a reality out of a dream fantasy like a long-range rocket.[1]

Dornberger and Wernher von Braun often talked about the bizarre twist of fate that had catapulted Hans Kammler into control of the V–2s, a subject about which the SS officer knew virtually nothing. They regarded him as a dynamic, indefatigable opportunist. On one day alone, Dornberger received 123 teletype orders from Kammler, many of them contradictory, most of them meaningless with regard to technically advancing the V–2.

Steadily, General Dornberger grew despondent and seriously considered feigning ill health and resigning his army commission. However, von Braun talked him out of the drastic action, pointing out that the V–2 would never be perfected without Dornberger's stellar leadership. So by holding his temper and proposing new ideas to Hans Kammler in a way that led the egotistical special commissioner to believe that he himself had originated them, Dornberger kept the missile program moving ahead.

In Moscow, Premier Joseph Stalin and other bigwigs in the Kremlin were aware of Germany's successful test-firings of the long-range V–2, a weapon that could revolutionize warfare if perfected. The Soviets had no missile research and development program, so in order to close the gap of perhaps 25 years, Stalin set up a special technical intelligence committee under Georgi Malenkov, chairman of the Council of People's Commissars. Armed with lists, swarms of Malenkov's agents were to enter the Third Reich on the heels of the advancing Red Army, fan out, and collar as many German rocket experts and installations as they could lay their hands on.

With Soviet tank-tipped spearheads charging across Poland toward the German frontier in July 1944, Malenkov's secret group, hoping to gain an enormous scientific edge over the Americans, British, and French, targeted Peenemünde for a cloak-and-dagger operation. Nine German prisoners of war, all of whom lived in the Peenemünde region, were given Reich money, false documents, and shortwave radios. They were parachuted into the Peenemünde vicinity on a dark night with the task of obtaining as much information as possible about German missile development.

As soon as the parachutists landed, all but one of them shucked his radio and began heading for their homes in the Fatherland, figuratively thumbing their noses at their Russian controllers. Only Lieutenant Erwin Brandt radioed back some brief items of interest. After his seventh message, Brandt was tracked down by the highly efficient *Funk Abwehr* (radio intelligence) and was arrested and executed.[2]

Meanwhile on the Western Front, Allied armies, bottled up in Normandy for six weeks, broke through tenacious German defenders, liberated Paris, and by mid-August were charging toward the German border. In total disarray, remnants of the once vaunted Wehrmacht were fleeing behind the Siegfried Line. Hitler's Third Reich appeared to be on the brink of total collapse.

In Washington, large numbers of army officers were assigned to plan for the remaining task–crushing Japan, a bloody struggle the strategists felt might go on for four more years. A study conducted by the Joint War Plans Committee of special weapons that could be used in the Pacific fighting concluded that missiles would be "highly desirable" as a means of destroying the industrial heart of Japan. The same report pointed out, however, that the development of America's only missile, the JB–2, was so retarded that its refinement in time to be used against the Japanese homeland was "highly unlikely."

In London, Winston Churchill was told by his military advisors that the advancing Soviet army was nearing the German missile testing center near Blizna, Poland. So the prime minister cabled Russian dictator Joseph Stalin and asked, in the interest of Allied cooperation, to permit a team of British and American technical experts to inspect Blizna once it was captured.[3]

Stalin promptly granted permission. British Colonel Terence Sanders and U.S. Lieutenant Colonel John O'Mara, along with an assortment of technical experts, flew to Moscow, then on to Poland. Reaching Poland on August 19, O'Mara and Sanders were told by Russian generals that Blizna still was being defended by German forces. Actually, the V–2 test center had been captured by the Russians two weeks earlier.

The joint British-American team was suspicious and angry. But all its members could do was to cool their heels for 12 days. Finally, the delegation was allowed to proceed to Blizna. On arrival, the mission soon discovered that the Soviets already had hauled away nearly everything of value, presumably including any V–2s. So the Brits and Americans packed rocket debris into crates, but when they opened them in London, scientists found that the haul consisted only of rusted airplane parts.[4]

Despite his cursory knowledge of missiles, General Hans Kammler, an architect by profession, was no fool. He knew that the Luftwaffe's buzz-bomb deluge on London had been launched from fixed sites along the English Channel in France. The launchers had been erected in sparsely populated areas, so the Allies had been able to use bomber fleets to plaster the sites, without fear of slaughtering countless numbers of civilians. Kammler's plan would provide no military target for Allied bombers to hit. Once a mobile V–2 battery would fire all of its missiles, it would pack up and move to another locale for the next firing.

A mobile V–2 battery consisted of three Meillerwagens, each carrying one rocket. A Meillerwagen was pulled by a prime mover, usually a half-track which also carried the firing crews. Then there were three tank cars, one with liquid oxygen for three missiles, one with alcohol for three missiles, and one for auxiliary fuels and other equipment. There was one generator truck for electric power and a truck that was the equivalent of a gun director, plus staff cars for the officers.

When the firing site was selected in Holland, the ''line of sight'' to London (about 200 miles away) would be established and three V–2s placed on firing tables. It would require about six minutes from lift-off for the rocket to soar to a height of about 50 miles and impact on or near London. By early September 1944, Hans Kammler was ready for the signal to commence firing.

In late August, war-weary Londoners were breathing sighs of relief. Surging American, British, and French armies on the Continent had overrun most of the fixed buzz-bomb sites along the Channel coast, so the nightmare of sudden death from flying robots seemed to be over.

But tension continued to grip the civilian population in England. For the past few weeks, Joseph Goebbels, Hitler's propaganda mouthpiece, had been trumpeting over Radio Berlin that the fuehrer was nearly ready to launch another frightening new secret weapon against Britain. An enormous new cannon, Goebbels explained, would hurl huge shells which, on explosion, would freeze to death any human within 200 yards of impact. The ''freezing shell'' was a figment of Goebbels' fertile mind, a psychological play to cast fear over the British homefront.

At 6:48 P.M. on September 8, while many Londoners were eating dinner, an enormous explosion rocked the suburb of Chiswick-on-Thames. Terrified civilians dashed for basements. The blast demolished 19 Chiswick houses and gouged out a 30-foot-deep crater in the ground. Scores of dead and injured were dug out of the ruins.[5]

British technical experts rushed to the site and interviewed civilians. From their descriptions, it was clear that the V–2 onslaught against London had been launched. Residents told of the enormous thunderclap followed by a rustling sound. Since the V–2 traveled faster than sound, the explosion was heard first and then the missile's approach.

Winston Churchill and his War Cabinet quickly clamped a muzzle on any mention in the press or on the BBC of the V–2 assault. It would be more than two months, after 64 missiles had struck London and its environs, before the British public—and the world—would learn details of the ghastly new secret weapons.

Wernher von Braun was at Peenemünde when he got word of the first V–2 strike against London. Associates were surprised that he reacted to the electrifying news with no sign of emotion and remained silent.

''Have you no comment about your rocket'' a friend asked. Von Braun pondered the question, then remarked soberly, ''Yes. It behaved perfectly but landed on the wrong planet.''[6]

In desperation, Winston Churchill and his War Cabinet decided to pinpoint the V–2 launching sites, presuming that they were like the ones for the buzz bombs, fixed installations. British intelligence reached the conclusion that the V–2s were being fired from Holland. But where in Holland? Aerial photos failed to disclose any ramps or other fixed launchers.

Across the Atlantic in the Pentagon, Colonel Gervais Trichel, the ambitious chief of the rocket branch of Army Ordnance, was poring over reports about the Germans' success with the V–2 and, like the Soviets, concluded that the Third Reich was 25 years ahead of the United States in missile development. If this enormous time spread were to be erased, Trichel was convinced, there was but one solution: capture Germany's long-range missiles and the scientists and engineers who had developed the rockets, along with technical documents related to the V–2.

Colonel Trichel, a West Pointer who held a master of science degree from the Massachusetts Institute of Technology and a doctorate in electrical engineering from the University of California, negotiated a contract with the General Electric Company. This top-secret deal was code-named Project Hermes. General Electric was to develop a program for the capture of German rocket scientists and the development of long-range missiles.

At 2:30 P.M. on Sunday, September 17, 1944, the vanguard of three Allied airborne divisions and a brigade, packed into 2,800 C–47 transport planes and 1,600 gliders, began bailing out and crash-landing along a salient 60 miles deep into Holland, stretching from Nijmegen in the south to Arnhem in the north. It was the opening strike in Operation Market-Garden, a bold plan to form a huge bridgehead in Holland, bolt across the Rhine River, and dash eastward to Berlin.

General Hans Kammler, the special commissioner for the V–2 program, feared that the powerful invading force would capture the Hague, in Holland, so he reacted swiftly and pulled back his mobile missile batteries to cities to the north where they took cover. Almost at one, the V–2 barrage against London ceased. The British rejoiced.

For ten days, bitter fighting raged in Holland, but the Allied thrust was halted at Arnhem. Hans Kammler then rushed his V–2 batteries back into the Hague, and the next day, explosions again were rocking London.

Now Adolf Hitler gave Kammler a second target—Antwerp—the finest port in Northwest Europe, through which were pouring the millions of tons of supplies needed by the Allied armies on the Western Front, which extended for 500 miles along the German frontier from Holland to the Swiss border.[7]

On October 30, Reichsmarschall Hermann Göring, the most pompous of the Nazi leaders, paid a visit to the Luftwaffe (west) side of Peenemünde to inspect refinements being engineered to improve the accuracy of the flying bomb. Göring (Fat Hermann to subordinates behind his back) had been a fighter-plane ace in World War I. Rumor had it among German generals that, with the present war going badly, he had become addicted to narcotics and often rambled in conversation. At his luxurious mansion Karinhall in the countryside north of Berlin, visitors reported that Göring had taken to wearing rouge, lipstick, and fingernail polish.

Be that as it may, on this day at Peenemünde, the leader of the Luftwaffe was clad magnificently with a light-gray uniform complete with a bevy of decorations, including the Iron Cross at his throat. His legs were encased in high red boots of soft Moroccan leather fitted with silver spurs. On his head was a visored white cap. Huge rings of diamonds and rubies were on each finger, and his hand clutched a field marshal's baton encrusted with diamonds and other precious jewels.

Once Göring's business was concluded at Peenemünde West, he passed over the invisible boundary to the ''enemy's camp'' in Peenemünde East to receive a briefing by General Walter Dornberger on the antiaircraft rocket code-named

Wasserfall. Much to the annoyance of Dornberger and other scientists, Göring was bored by it all. Periodically, he opened a small bottle and popped a pill into his mouth.

Morale at Peenemünde had nearly hit bottom. On December 12, 1944, a *Volkssturm* (People's Army) unit was formed at the experimental center on orders from Berlin. Despite their customary 12- to 16-hour workdays, Wernher von Braun and other technical experts were trained as soldiers three times a week. For four hours at each drill session, some of the world's most brilliant scientific minds were taught hand-to-hand combat and street fighting.

A climate of foreboding gripped Peenemünde. Nine months after Reichsfuehrer Himmler had thrown von Braun and two other scientists into jail on trumped-up charges, Gestapo agents, disguised as workers and rotating assignments so as not to attract attention, were still watching every step the three rocketeers made. Von Braun, Klaus Riedel, and Helmut Grötrupp remained "enemies of the state."

Despite the stultifying climate at the experimental center on the Baltic, von Braun and his team were working to develop a multistage missile (code-named America) that could hit New York, Washington, Philadelphia, and other cities on the eastern seaboard or, as von Braun hoped, could boost a satellite into orbit around the earth.

Dubbed the A–9, the embryo missile would have a pressurized cockpit in place of the warhead to carry a pilot. Von Braun envisioned developing an A–10 booster rocket with a 440,000-pound thrust. If the manned A–9 were mounted as a second stage on top of the powerful A–10 booster rocket, it could become a supersonic rocket plane capable of crossing the Atlantic Ocean.

Four thousand miles west of Peenemünde on the bitterly cold morning of November 29, 1944, two German spies, William Colepaugh and Erich Gimpel, were passengers in a U-boat, embarked on *Unternehmen Elster* (Operation Magpie). They had been sent to the United States by the SS to ferret out technical information on rocket development and airplanes.

Shortly after noon, Colepaugh and Gimpel climbed from the submarine into a rubber raft off bleak Crabtree Point, Maine, and were paddled ashore by two sailors. The spies caught a train for New York City, where, after blowing most of the $60,000 given to them by two SS colonels, they were captured by agents of the Federal Bureau of Investigation (FBI).

Grilled by the FBI, Colepaugh and Gimpel swore that before leaving on their mission, the SS colonels had told them that a rocket attack on New York City would be launched soon. An intense probe by U.S. Naval intelligence produced vague evidence that U-boats had been equipped with "special devices" that would permit missiles and buzz bombs to be launched against U.S. coastal cities from over the horizon, and that the Germans had developed an enormous missile that could cross the Atlantic. The American investigators were unaware that the cross-Atlantic multistage missile was only in the development stage.

On the gray morning of January 8, 1945, a flock of reporters, pencils and notepads at the ready, hovered around Admiral Jonas H. Ingram, commander of the Eastern Sea Frontier, in his wardroom aboard a warship in New York harbor. The scribes had been promised an "historic press conference."

Ingram, a flat-nosed, heavyset old salt who had gained recognition as football coach at the Naval Academy, was one of the Navy's colorful characters—and most outspoken. Seated behind a long table, Ingram said: "Gentlemen, I have reason to assume that the Nazis are getting ready to launch a strategic attack on New York City and Washington with missiles or robot bombs."[8] There was a gasp of astonishment from the hard-bitten reporters.

"I am here to tell you that attacks are not only possible, but probable as well, and that the East Coast is likely to be hit within the next thirty to sixty days." Ingram eyed his listeners, then added grimly: "We're ready for them. The next alert you get is likely to be the McCoy. The thing is not to get excited about it. It might knock out a high building or two. It might create a fire hazard. It would certainly cause casualties. But it could not seriously affect the progress of the war."

Again the Old Salt paused. "But think," he added, "what it would mean to Dr. Goebbels at this stage of the war to announce that 'today we have destroyed New York City!' It would be very good politics for him."[9]

Within hours of the sobering press conference, New York City was abuzz with lively discussion about Admiral Ingram's warning. There was talk of missiles, rockets, buzz bombs, and V–3s (the cross-Atlantic missile). Jangled nerves were given another jolt a day later when the *London Daily Express* published a front-page story declaring that the Germans would "V-bomb" the United States in order to show off their new terror weapon for "malice and vanity" and to help their hard-pressed confederates, the Japanese.[10]

Meanwhile in Europe, Adolf Hitler's final roll of the dice to snatch victory from the jaws of defeat had resulted in a crushing disaster. On December 16, 1944, the fuehrer launched Operation *Wacht am Rhein* (Watch on the Rhine), a mammoth surprise offensive against American forces in Belgium in what came to be known as the Battle of the Bulge. After a savage pitched battle that raged for six weeks in the bitter cold, remnants of the attacking Wehrmacht limped back to positions in the Siegfried Line, from which they had leaped with such high hopes.[11] Now, even the staunchest German knew that the Third Reich was doomed. The lone exception to that viewpoint was Adolf Hitler.

# 9

## "Germany Has Lost the War!"

By the end of January 1945, Soviet armies, advancing from the east, were only 75 miles from Peenemünde. Two weeks earlier, the center's commandant, General Walter Dornberger, had been called to Berlin for a special assignment; Wernher von Braun, in essence, became the Peenemünde commander while retaining his post as chief technical advisor.

Now von Braun received orders from the Oberkommando der Wehrmacht to destroy everything that could not be evacuated and to make certain that the Soviets did not obtain any information that might permit them to reconstruct a V–2. Documents and drawings were prepared for destruction by placing containers of acid and visol in such a manner that when released they would cover the research data and upon mixing would cause combustion. Discussions were held about blowing up the test stands and the electrical power plant serving the Luftwaffe and army installations.

At this point, von Braun, at great personal risk, called a secret meeting of his top scientists and engineers at a farmhouse near Peenemünde. As the group convened, the muffled crump of exploding Soviet artillery shells could be heard far in the distance. Fear hovered ominously over the meeting, for the Gestapo might have gotten wind of it and could burst into the farmhouse at any moment.

"Germany has lost the war," von Braun declared. "But let us not forget that it was our team that first succeeded in reaching outer space. We have never stopped believing in satellites, voyages to the moon and interplanetary travel. We have suffered many hardships due to our faith in the great peacetime future of the rocket." Von Braun glanced around the group of grim, silent men, each immersed in his own thoughts. "Now we have an obligation," he continued. "Each of the conquerors [Russia and the Western Allies] will want our knowledge. The question we must answer is: to what country shall we entrust our heritage?"[1] There was virtually no debate and the verdict was unanimous: the missile scientists and engineers would try to surrender to the Americans.

Germany was disintegrating rapidly. Chaos was rampant. The mass confusion spilled over into Peenemünde where von Braun, in the absence of General Dornberger, received a flood of contradictory orders from overlapping military commands and governmental bureaus.

One order was from the army general in command of the region: von Braun and the scientific experts were to lay aside their tools, pick up their rifles, and defend Peenemünde to the last man.[2]

Hard on the heels of that chilling order came one from the Armament Ministry in Berlin. Von Braun, his scientists, engineers, and their families were to move with most of the important research documents and equipment to Bleicherode, a town near the Harz Mountains, where they were to continue efforts to boost the range of the V–2 and make it more accurate.[3]

Meanwhile, the Allied air barons in London also were focusing on the Harz Mountains and the underground assembly plant. That facility had become a menacing threat for it was mass-producing V–2s and revolutionary new jet-airplane engines. If the Germans could manufacture sufficient jet airplanes, they could conceivably drive the Allies' powerful, much slower, propeller-driven air forces from the skies. And the V–2 potential remained a frightening specter.

The air barons concluded that the protective nature of the underground factory would shield it against even a deluge of 2,000-pound bombs. So General Carl A. "Tooey" Spaatz, commander of the U.S. Strategic Air Forces in Europe, proposed a novel scheme to wipe out the plant. Bombers would discharge over the Harz Mountains thousands of gallons of a mixture of gasoline and soap, which would seep in through the ventilation system, then burst into a raging inferno of fire in the underground chambers. At the same time, fighter bombers would plaster all exits with explosives, sealing the factory and preventing people from getting out or assistance from getting in.

After a prolonged discussion, the idea was dropped. It would have resulted in the hideous deaths of thousands of innocent slave laborers and civilian workers.

Back in the United States early in 1945, the energetic ordnance officer, Colonel Gervais Trichel, scrounged sufficient funds from the Pentagon to establish a U.S. Army missile firing range in a bleak locale in New Mexico. It was named the White Sands Proving Ground. If the United States were to close the 20- to 25-year gap in scientific missile know-how (and save tens of millions of dollars), it would be most beneficial to the General Electric engineers involved in Project Hermes to test-fire V–2 missiles at White Sands.

So in early March 1945, Colonel Trichel rushed an urgent request to Colonel Holger N. "Ludy" Toftoy, chief of Ordnance Technical Intelligence in Paris. Trichel's shopping list was for 100 V–2s in good operating condition to be shipped to the White Sands site in New Mexico at the earliest possible date.

West Pointer Toftoy, a can-do type, could not believe his eyes. Although his job was to locate captured German weapons and send them back to the United States for evaluation by experts, how was he supposed to "capture" 100 46-foot missiles, each weighing 13 tons, from under the noses of the Germans? If he had a ghost of a chance of filling Gervais Trichel's shopping list, photo reconnaissance and interrogation of German prisoners indicated that the most logical source of supply was the underground factory in the Harz Mountains.

The other U.S. military services also had leaped into the hunt for German scientific booty. By April 1, 1945, there were 14 technical intelligence teams from the Army, Navy, and Air Corps, each independent of one another, skedaddling around western Germany to sniff out what they could, often competing with each other—and with their Soviet, British, and French counterparts. It had the trappings of a colossal treasure hunt, with gigantic and far-reaching rewards going to the victor.

Admiral Ernest J. King, the hard-boiled U.S. chief of naval operations, had dispatched to Europe the largest and most heavily financed investigative team. Under Commander Henry Schade, King's Naval Technical Mission, 236 strong, dashed about in a wide assortment of conveyances—including a private airplane for Commander Schade. The Navy officer's aircraft was the envy of his American competitors, for they had to hitchhike when their work required a flight.

General Henry H. "Hap" Arnold, commander of the U.S. Army Air Corps, had formed a top-secret project called the Army Air Forces Scientific Advisory Group, which had been prowling France and Belgium and now Germany in the wake of the advancing American armies. Led by the American rocket pioneer, Theodore von Karman, the air group already had uncovered a mind-boggling array of revolutionary German aerodynamic developments.

Among those deeply impressed by German scientific advances was Major General Hugh J. Knerr, deputy commanding general for administration of the U.S. Strategic Air Forces in Europe. Knerr sent a detailed memorandum to his boss, General Carl Spaatz:

> Occupation of German scientific and industrial establishments has revealed the fact that we have been alarmingly backward in many fields of research. If we do not take this opportunity to seize the apparatus and the brains that developed it and put the combination to work promptly, we will remain years behind while we attempt to cover a field already exploited. Pride and face-saving have no place in national security.[4]

The wide-ranging, free-wheeling search for German scientists and their secrets was spurred by intelligence reports confirming colossal German advantages in an entire range of weaponry on land, sea, and air. In addition to the most coveted prize of all—the V–2 missile—there were other spectacular German technological advances that far exceeded anything the Allies possessed and awaited the most enterprising of the invading nations.

Revolutionary jet airplanes were described as having "exceptional performance," and the Luftwaffe had radio-controlled Hs 239 and FX bombs for which the Allies had no counterparts. German infrared receivers reportedly could detect airplane exhaust fumes from a distance of 12 miles.

Allied intelligence reports revealed that two new types of submarines, both extremely swift, virtually undetectable, and able to recharge their batteries while under water, were under construction. Some of these submarines used a new method for firing rockets while submerged.[5]

Richard Porter, the young genius who led the Project Hermes team for Army Ordnance, had guided-missile targets to investigate other than the V–2s, a task that kept him hopping all over western Germany. The Germans had developed a series of revolutionary antiaircraft missiles, including the Wasserfall, *Rheintochter* (Rhine Maiden), *Enzian* (Gentian), *Taifun* (Typhoon), and *Schmetterling* (Butterfly).

Front-runner in the early stages of the Allied search for German long-range missile secrets was Great Britain, whose Lord Cherwell, Alwyn Crow, and other scientists had long been involved in rocket experimentation. Duncan Sandys' Bodyline Committee; its successor, the Crossbow Committee; MI–6; and other intelligence agencies for years had been ferreting out the identity of the German missile scientists and engineers, as well as V–2 installations.

The British lead steadily eroded, however, in the wake of stiff competition with the Americans. Field Marshal Bernard L. Montgomery, the peppery commander of the British 21st Army Group, was among those who seemed to be indifferent to beating the American scientific competitors. He assigned the investigative function to his chemical warfare team. Its members, the diminutive field marshal declared, were "very enthusiastic and active in embarking on numerous treasure hunts." They also were unqualified for the task.

Allied scientific investigators had their job rendered far more difficult by a personal order that Adolf Hitler, in a fit of pique, had issued to his commands when the Third Reich's doom became vividly apparent after General Eisenhower's forces stormed across the Rhine. On the approach of enemy spearheads, the fuehrer ordained, all scientific research and development facilities, installations, and their documents were to be destroyed.

Albert Speer, the armaments chief and one of the few men in Germany who dared to speak up to the fuehrer, tried to reason with him. Red-faced with fury, Hitler shouted back: "If the war is lost, the nation will also perish. It will be better to destroy things ourselves because this nation will have proved to be the weaker one and the future will belong solely to the stronger nations."[6]

# 10

# Journey to the Bavarian Alps

Adolf Hitler's Third Reich was shrinking under the Allies' pressure from all sides, yet in his Chancellery, the fuehrer was breathing defiance. Over Radio Berlin, on the tenth anniversary of compulsory military service under the swastika, he alluded to his personal martyrdom and concluded: "Our task is clear: to put up resistance and to wear down our enemies, so that, in the end, they will tire of the war and will yet be defeated."[1]

Meanwhile, Hitler continued to scream at his commanders to destroy the Ludendorff Bridge over the Rhine River at Remagen. On March 16, 1945, elements of the U.S. 9th Armored Division had seized the bridge when German engineers failed in an effort to blow it up as ordered. American tanks and troops poured across the span and within a week had carved out a bridgehead ten miles deep and 23 miles long.

In his desperation, the fuehrer ordered that his supersonic V–2s be turned against the Ludendorff Bridge. For the first time in history, long-range missiles would be employed in a tactical capacity on the battlefield.

Learning shortly in advance of the intention to "wipe out" the Ludendorff Bridge with V–2s, several Wehrmacht battle leaders in the Remagen region mildly protested to their military superiors, citing the peril from the relatively inaccurate weapons to thousands of German civilians. When the objections went up the pecking order to the fuehrer, he brushed them aside, stating that he intended to use his missiles to halt Americans at the Rhine.

With typical alacrity, General Hans Kammler fired a total of 11 V–2s from mobile launchers in Holland. Three landed in the Rhine River a short distance from the Ludendorff Bridge, five others exploded west of the span, and another, its guidance mechanism gone awry, crashed west of Cologne, 40 miles up the river.

Two days after the seizure of the Ludendorff Bridge, elements of the U.S. 1st Infantry Division (the Big Red One) fought their way into Bonn, 20 miles south of Cologne on the west bank of the Rhine. In peacetime, the city had been noted chiefly as the seat of famed Bonn University and the birthplace of Ludwig von Beethoven. Founded by the ancient Romans as a fortified encampment, it had been known as *Bonna* or *Castra Bonnensia*, and there Julius Caesar's soldiers had maintained their position on the Rhine against restless German tribes on the east bank.

Now, Bonn had been pounded heavily by American artillery and bombs, and more than half of the city's buildings had been damaged. With the approach of spearheads of the 1st Infantry Division, professors and scientists at Bonn University were ordered to destroy all documents, blueprints, and other materials with regard to new weapons design, upon which they had been working for the Wehrmacht. In their haste, the top-secret papers were torn up and flushed down toilets at the university. However, one toilet had not flushed properly and a Polish technician, who had "volunteered" to work at the facility, fished out the shredded strips of paper and later handed them over to Allied intelligence.

The unheralded and anonymous Pole unwittingly had scored an intelligence bonanza for the Allies. When the scraps of paper were dried out and tediously pieced together, it was found that the sheets included a compilation of all scientific projects and of the scientists and engineers who had worked on them during the war. This priceless document was dubbed the Osenberg List, for it had been laboriously compiled over many years by Fritz Osenberg, who had been in charge of the Planning Office of the Reich's Research Council.

Although British intelligence had gotten its hands on the Osenberg List first, the list soon reached the desk of U.S. Army Major Robert Staver in a small office in London's Grosvenor Square. Staver, who graduated from Stanford University in 1940 with a degree in engineering, had been in the forefront of rocket development in Army Ordnance for the past three years. Outgoing and eager, the 28-year-old major had been sent by Colonel Trichel in Washington to locate and interrogate German rocket scientists and engineers, as well as to secure documents and blueprints concerning the V–2.

Staver's task was a daunting one. From the time of his arrival in London in mid-February 1945, he put in 12- to 14-hour days, seven days per week, to compile detailed lists of hundreds of German installations all over Europe where missile and jet-propelled work was being done. He then began compiling lists with the names of technical experts who were involved at those sites.

Staver could not have done the job without the wholehearted cooperation of British rocket experts. They not only provided the American major with the Osenberg List but also turned over to him voluminous files that had been collected during the Peenemünde investigations. It was from a detailed Target Information Sheet, which included scores of aerial photographs, that Major Staver knew that the Harz Mountains were where the V–2s were rolling off assembly lines.

Then Staver began the arduous task of giving priority to the thousands of names of technical experts who could yield information of significant military importance or who were a potential threat should they fall into threatening hands (the Soviets). Staver's collection was given the code name Black List, at the top of which was the name Wernher von Braun.

On the night of March 23–24, 1945, Allied armies in the West launched Operation Plunder, the assault crossing of the majestic Rhine River, which had been significant in German history ever since Julius Caesar built a timber bridge across it more than 2,000 years earlier. Reinforced after daybreak by history's largest simultaneous airborne assault by the U.S. 17th Airborne and British 6th Airborne divisions, the attackers quickly got a foothold on the far shore.

Adolf Hitler's reaction to news that the Allies were over the Rhine in strength was towering rage. He promptly ordered that the thousands of American and British airmen who were prisoners of war in the Third Reich be shot. Cooler heads prevailed.

The fuehrer also demanded that poison gas be unleashed against the enemy in the West. Unknown to the Allies, scientists with the German industrial giant I. G. Farben covertly had created a revolutionary lethal nerve gas (Tabun), against which neither Eisenhower's armed forces nor the civilians in England had any protection. If this gas were loaded onto a V–2, it could have wiped out the entire population of London.

I. G. Farben had code-named its lethal gases Trilon (the name of an ordinary detergent). Under orders from the Oberkommando der Wehrmacht, Otto Ambros, a chemist and official in I. G. Farben, had supervised the top-secret construction of a factory in Breslau, Silesia, for production of Tabun, which laboratory tests on animals proved could kill a human in from five to 15 minutes.

In February 1945, Hitler ordered the collections of Tabun at the secret factory in Silesia to be loaded onto barges on the Elbe and Danube in order to launch a final V–2 attack that he hoped would massacre millions of British civilians.[2]

In the meantime, after crossing the Rhine, American spearheads were charging eastward in the direction of the Harz Mountains. In the south, other U.S. task forces were driving rapidly toward the Bavarian Alps, where, Allied intelligence had (erroneously) reported, Adolf Hitler and his top leaders would gather for a last-ditch stand by 250,000 die-hard SS troops in what Joseph Goebbels called the Alpine Redoubt. British forces were heading toward the north and northeast and overran much of Holland.

Although no one in the Allied camp was aware of it at the time, the final V–2 exploded in London on March 27, 1945, killing 127 civilians and wounding a few hundred others.

Yet Hitler clung to the hope that his V–2s and other secret weapons could turn the tide against the Allies. He ordered Hans Kammler to collect all the V–2s from Holland and rush with his mobile firing batteries into the Harz Mountains. On arrival, the fuehrer directed, Kammler was to organize a fortress, defend it to the last man with 80,000 elite SS troops, and resume pounding London with the missiles.

If Hans Kammler needed his resolve to be steeled, he had only to listen to

Radio Berlin. Over the airwaves, diminutive, 47-year-old Joseph Goebbels screamed defiance: "Berlin and all of the Fatherland will be defended stone by stone!" Robert Ley, Hitler's labor chief, exhorted Germans: "We will fight in front of Berlin, in Berlin, behind Berlin!" Pouring out of Radio Berlin night and day was a constant barrage of exhortations to "true" Germans to resist the invaders, together with threats to "traitors, shirkers, and defeatists."

With his customary boundless energy, Hans Kammler began to carry out his orders, positioning the SS units as they arrived in the Harz Mountains and clearing launch sites on a high plateau inside the fortress. He ran into enormous difficulties, however, while trying to get the missiles into place to fire on London. And American spearheads were only 50 miles to the west—and closing in fast.

Meanwhile at Peenemünde, a curious conglomerate of conveyances began the mass movement of 5,000 V–2 technicians and their families toward Bleicherode in the Harz Mountains, 250 miles to the southwest. A few hundred trucks, many of them puffing and wheezing due to age and lack of maintenance, carried people, and two trains were loaded with tons of scientific equipment and precious documents.

The long trek would be fraught with peril. Swarms of Allied fighter-bombers crisscrossed the skies and pounced on any German vehicle on the roads, making it necessary to limit travel to the nighttime hours. Bridges had been knocked out. Road signs had been removed (to confuse the advancing Allies).

Moreover, the route took the convoy directly through the rear area of the Eastern Front commanded by the German general who had ordered von Braun and the others to fight to the last man to defend Peenemünde against the Soviets. Bands of SS troops and Gestapo agents roamed the area in search of army deserters and civilians who could not produce official orders accounting for their movements. Von Braun had no written orders.

About halfway to Bleicherode, Wernher von Braun's car, at the head of the caravan, was halted by a German army major. Civilian vehicles were not allowed in the area, the major declared sternly, so the convoy would have to turn back. It was an anxious moment. Retracing their route could have resulted in capture of the rocketeers and their precious cargo by the advancing Russians.

Coolly, von Braun assured the major that his convoy was involved in a top-secret project that was vital to the war effort. It had been ordered to proceed to Bleicherode by Reichsfuehrer Himmler personally, von Braun lied. The major's face twitched at the mention of Himmler: even lower-ranking officers feared the boss of the SS and the Gestapo.

Authenticity was lent to von Braun's hoax by the fact that each truck, automobile, and other vehicle in the convoy had blazing red-and-white signs with the fictitious designation *VZBZ*. The scientist explained to the major that this meant *Vorhaben zur Besonderen Verwendung* (project for Special Dispositions). Von Braun's caravan was waved forward.

Six weeks were required for the entire movement to Bleicherode. Hounded by Allied warplanes, bombed, strafed, and stalled before wrecked bridges, von

Braun found that no preparations had been made for the arrival of the Peenemünde group, which included children. So he issued official-looking but phony orders, based on the V–2 priority status, and requisitioned lodging for families in surrounding villages.

The torturous odyssey through war-ravaged Germany convinced von Braun that the situation was hopeless. Yet, with typical zeal and drive, he set about carrying out his orders from the Ministry of Armaments to improve the accuracy of the V–2. Secretly, he hoped to use the remaining days of the conflict to conduct experiments that might advance postwar space travel and interplanetary exploration.

Von Braun's work required that he travel to Berlin on occasion. Always, these trips had to be made at night because of the constant threat of American fighter-bombers. On the morning of March 16, 1945, von Braun left Bleicherode for the 60-mile dash to Berlin at 2:00 A.M. His young civilian chauffeur at the wheel of the Hannonag Storm got on the *autobahn* (four-lane divided highway) at Naumberg and sped along with dimmed lights.

Von Braun, in the meantime, settled back to grab a few winks of sleep before arriving at the Ministry of Munitions; he hoped to arrive before dawn turned the autobahn into an Allied shooting gallery. Halfway to Berlin, the driver dozed off, and the Storm swerved over a railroad embankment and plunged into a ditch.

Badly injured, the scientist forced open a door, wriggled out, tugged the unconscious chauffeur from behind the steering wheel, and dragged him away from the wreckage for perhaps 15 yards. Moments later, the Storm burst into flames. Then von Braun passed out.

When von Braun regained consciousness, he was in a hospital. His face had been cut, his left shoulder shattered, and his left arm broken in two places. The chauffeur had a fractured skull, but survived.[3]

Five days after his brush with death, on March 21, von Braun talked his doctors into releasing him from the medical facility, knowing that his people in Bleicherode needed his leadership in this chaotic period. Two days later, with his arm and chest encased in a plaster cast, von Braun was honored by Walter Dornberger and a few associates with a small party celebrating his 33rd birthday. It was a subdued affair. Everyone knew that the days of the Third Reich were numbered.

As April approached, even the resolute General Hans Kammler reached an identical conclusion: the Third Reich was indeed *kaputt* (finished). Although he would continue to fight, his primary concern now was saving his own skin, for the victorious Allies might charge him with war crimes for the often brutal handling of thousands of slave laborers and political prisoners working in the Mittelwerk in the Harz Mountains.

Both the Western Allies and the Soviets would be deeply interested in learning the secrets of the V–2, Kammler astutely realized. Like Wernher von Braun and his rocketeers, the SS general concluded that his greatest hope for favorable treatment (or even survival) lay in falling into American hands.

Kammler knew, however, that he himself would be of little value to the Americans because of his minuscule knowledge of guided-missile technology. So the SS zealot hatched a scheme whereby he would hold his 500 rocket experts as hostages to use as bargaining chips with the Allies. Von Braun and the others would be captives until Kammler deemed that it was the right moment to surrender.

On April 1, Kammler called in Wernher von Braun. No time was wasted on idle chit-chat. "You will select 500 of your key men," the general declared, "and have them ready to embark on my private train tomorrow. No families will be permitted. I'm sending you where you can continue your important work without fear of being captured."[4]

There were gut-wrenching scenes at the Nordhausen railroad station on the following afternoon. Families bid tearful farewells, wondering if the members would ever see one another again. Then the 500 scientists, engineers, and technicians, some wiping tears from their eyes, climbed aboard Kammler's train, complete with its dining and sleeping cars. Just before dusk, the train chugged out of the station and headed southward for the Bavarian Alps, 400 miles away.

Von Braun and a few of his men remained behind for a few days to take care of some crucial business. Personal orders had come from the fuehrer to make certain that V–2 documents and other data did not fall into the hands of the Allies. Then von Braun and the others would follow by car to the Bavarian Alps.

On April 3, Adolf Hitler summoned General Kammler to a meeting in the *Fuehrerbunker* below the bomb-battered Chancellery in Berlin. The dashing, supremely confident young SS general had become a Hitler favorite, surpassing Albert Speer in the inner circle. The session continued far into the night.

A day later, Propaganda Minister Joseph Goebbels scrawled in his personal diary: "SS Obergruppenfuehrer Kammler now shoulders the major responsibility for [building new jet warplanes] as well as the V–2 rocket program." Kammler, Goebbels added, behaved "in splendid fashion" and Hitler "pinned great hopes on him."[5]

While the fuehrer was pinning his "great hopes" on the opportunistic General Kammler, the Reich rocket boss was scheming to save his own hide in the looming collapse of Germany. Although Kammler and Albert Speer detested one another, the SS general summoned the armaments' chief to his Berlin office after leaving the Fuehrerbunker.

Kammler bluntly got right to the point with Speer. The war had been lost, the SS general declared, and Speer would have a better chance of survival if he were to leave Berlin at once and join Kammler in the Bavarian Alps. Speer was shocked when Kammler stated outright that there were "moves afoot" to dethrone the fuehrer, and that if they were successful, Kammler planned to contact the Western Allies (the United States and Great Britain) immediately and offer them the secrets of the V–2 rocket and the turbojet in exchange for personal freedom.[6]

Only much later would Albert Speer learn what Kammler meant by ''moves afoot'' to rid Germany of the fuehrer. Kammler's mentor, Reichsfuehrer Heinrich Himmler, naively perhaps, had been in secret contact with the Western Allies as long ago as November 1944, with the suggestion that he, Himmler, take over as head of the German nation if Hitler could be ''removed'' from power. Speer turned down Kammler's invitation to flee southward to Bavaria.

Rain was beating down on the night of April 4 as a decrepit truck, driven by wood gas, whose windshield wipers and brakes refused to function adequately, was huffing along a secondary road that led from Bleicherode to the small town of Dörnten. Along with the harried driver, who had to halt every mile or so to scrape rain and mud from the windshield, were two of von Braun's engineers, Bernhard Tessman and Dieter Huzel. A squad of SS troops, under a captain, rode in the cabs of two other trucks.

Under a heavy canvas cover in the bed of the first truck were perhaps three tons of valuable documents related to the development of the V–2. Other papers and materials, weighing perhaps 11 tons, were being carried in the other trucks in the convoy.

Now the little caravan edged its way through blacked-out Dörnten and then onward to an abandoned iron ore mine just outside of town. There men in the trucks dismounted and began shouldering cartons of the documents and stashing them away inside the black, damp mine. After the entrance was sealed shut by blasting, the Germans hurried away.[7]

A day later, von Braun drove to the Bavarian Alps where he was assigned comfortable quarters in the quaint, beautiful village of Oberammergau in the foothills some 40 miles south of Munich. Known over the centuries for its skilled wood carvers, Oberammergau conducted its first Passion Play in 1634, with a solemn pledge to present it every ten years if the Black Plague would cease. Even during the World War II years, that vow continued to be fulfilled.

Almost at once von Braun drove to where his 500 men were assigned to an empty army camp just outside of Oberammergau. There he was jolted. Barbed wire and SS guards surrounded the group of barracks. His men were prisoners.

Meanwhile, General Hans Kammler had moved his headquarters from Berlin to Bavaria and ensconsed himself in the Haus Jesu Christi Hotel in Oberammergau. The favorite of Adolf Hitler and Heinrich Himmler occupied the best suite in the biblical structure, and the innkeeper and staff bent over backwards to provide their VIP with service and whatever extras could be scraped up.

A day after Wernher von Braun's arrival, General Kammler sent for the rocket scientist. While waiting in an adjoining room to see Kammler, von Braun overheard a conversation between the SS general and his trusted aide, SS *Obersturmbannfuehrer* (Lieutenant Colonel) Ludwig Starck. Von Braun was shocked by the conversation: the two men were talking about escaping possible American capture by holing up in the nearby 14th-century Ettal Abbey and donning the garb of the Benedictine Order, famous worldwide for the liqueur made by the monks.[8] Starck proposed that the two men burn their SS uniforms and remarked, perhaps half in jest, that they might find postwar jobs as liqueur salesmen.

Von Braun heard footsteps in the hall and quickly moved himself—and his ear—from the wall where he had been eavesdropping. Colonel Starck entered the room and escorted von Braun to General Kammler's suite. No sooner had the rocket genius taken his seat than he noticed Starck standing by with a machine pistol resting on a table at his side. Clearly, Kammler trusted no one at this chaotic stage in the Third Reich.

Calm and collected as always, Hans Kammler inquired if the Peenemünde rocket men were comfortable in their new setting and if they would be able to resume rocket research. They could do much better, von Braun replied, if the rocketeers were moved from their SS barracks, where Allied bombers might wipe them out, to villages scattered around the region. Kammler quickly gave his approval to the dispersal of the experts.

A few days later, the 500 missile men, escorted by SS guards, were dispersed into billets in 20 villages in the area, where they would not all be killed by a single bombing raid. Von Braun himself settled in Weilheim, 20 miles south of Oberammergau, where his 26-year-old brother Magnus joined him. The von Braun boys worried deeply over the unknown fate of their parents, the baron and baroness, who were living in the region overrun by the Red Army.[9]

Not long afterward, General Walter Dornberger, with a small military staff, also joined the "Peenemünde 500," establishing a command post in the Bavarian ski resort of Oberjoch.

In recent days, Wernher von Braun's arm and shoulder, injured in the auto accident, were growing steadily more painful. Fearing that he might have to have the arm amputated if he did not receive medical attention, he checked into a hospital in the nearby village of Sonthofen, where physicians specialized in setting the broken bones of skiers.

Von Braun's arm had to be broken once more and reset. No anesthetics were available, so the procedure resulted in excruciating pain. Then the patient, in agony, was put into traction and the surgeon told him that a second operation would be necessary in a few days.

In his basement room, von Braun slept fitfully. His pain was intense, he was worried about his parents and his 500 men, and Allied planes seemed to be overhead almost constantly. One night, he awoke with a start and saw a uniformed man stooping over his bed. Von Braun's first thought was that a Gestapo agent had come to murder him to keep him out of the hands of the Allies. Then he became aware that the dim figure was wearing a German army uniform. The stranger held a finger over his lips to signify the need for silence. "General Dornberger sent me to bring you out," the soldier whispered. "He says you must act now." A surgeon was summoned and he cautioned the scientist (whom he did not recognize) against leaving. "You might lose your arm," he exclaimed.

Finally, the physician agreed to apply a new cast to von Braun's chest and left arm. Moments later—the pain still agonizing—the scientist and the soldier stole out of the hospital and into a waiting ambulance that General Dornberger had somehow "requisitioned."

With the soldier at the wheel, the ambulance raced through several villages and came to a halt before a small hotel in Oberjoch. Von Braun quickly recognized the figure that emerged and walked swiftly toward him. Old friends Walter Dornberger and Wernher von Braun were reunited.

Huddled in a room in the three-story hotel, von Braun now told his confidant about his plan to surrender the Peenemünde research and development team to the Americans. "Do you think I would have sent for you, were I not convinced that this is the right course to take?" Dornberger replied. "The war is over. Now it is our obligation to mankind to place our baby [the V–2] in the right hands."[10]

Von Braun broke out his flashing smile.

While the two friends awaited developments, they took long walks in the peaceful mountains, played chess, worried about their relatives and friends, and talked of future space travel. Wernher von Braun was especially enthused with a visionary proposal that he had developed along with his former mentor, Hermann Oberth.

Their scheme was to one day build an "observation station" in outer space. The erection would be easy, von Braun concluded. Components would be sent into the interstellar spaces by means of rockets; the components would have no weight in the state of free gravitation. "The work would be done by men who would float in space, wearing diver's suits, and who could move at will in space by means of small rocket propulsion units, the nozzles of which they would point in the required direction," von Braun said enthusiastically.

This observation station could be equipped with an enormous mirror, consisting of a huge net of steel wire onto which thin metal foils would be extended. A mirror of this type would have a diameter of many yards, and its component facets would be controlled by the station, which would enable the heat and light of the sun to be concentrated on selected points of the earth's surface. This would enable large cities to get sunlight during night hours of darkness.

What was more, von Braun said, "The weather, too, could be influenced by systematic concentration of the sun's rays onto selected regions of the earth." Droughts could be abolished by inducing rain to fall on certain areas, by concentrating the sun's rays onto distant lakes and seas, thereby increasing their evaporation. The clouds thus formed could be driven to the required locale by influencing the centers of low and high pressure through radiation from other facets of the mirror.

Although Walter Dornberger was enjoying his relaxed interlude in pristine Oberjoch, his mind was ill at ease. Reports had reached him that the British blamed him for the V–2 assaults on London and were anxious to give him "a fair trial and hanging."

Far north of peaceful Oberjoch, the *Kriegsmarine* (Navy) and die-hard SS officers had concocted a scheme for continuing to heap death and destruction on England's civilian population, even as the Third Reich was crumbling. Plans were to build submergible barges on which would be loaded V–2 missiles ready for launching. These barges would be towed into the North Sea by manned, submerged submarines, and the launchings would take place after the barge surfaced.

Workers at the Vulkan Shipyards in Stettin already had built three of the unique submergible barges when the city was overrun by the Soviet army.

# 11

# Discovery at Kohnstein Mountain

Some 260 miles southwest of Peenemünde, Lieutenant Colonel Andrew Barr, a Securities and Exchange Commission official in civilian life and now G–2 (intelligence officer) of the U.S. 3rd Armored Division, was seated in his "office," a half-track, complete with a built-in desk with map drawers underneath and a small stand for a typewriter. Barr was in a column of the 3rd Armored that was rapidly approaching the Harz Mountains. It was April 10, 1945.

Barr was puzzled by the strange scraps of information that he and his staff had been collecting within the past 24 hours. Not knowing what to make of the situation, Barr sent out by wireless a series of "alerts" to the division, whose advancing columns were strung out for many miles.[1]

One of those perplexed by the nature of the "alerts" was young Lieutenant Colonel William A. Castille, an intelligence officer of Combat Command B of the 3rd Armored Division. While interrogating captured German soldiers, Castille was told by Barr that any mention by the enemy of unusual factories, strange freight cargo, "heavy water," and the like was to be flashed back immediately to higher headquarters.

"What kind of 'unusual factories' and 'strange cargo' do you mean?" Castille radioed Colonel Barr. "Just what I said—*any* unusual or strange things you might hear about," Barr replied.[2]

Bill Castille reflected that Colonel Barr might have seen too much combat action. Since the Allied breakout from Normandy in July 1944, Major General Maurice Rose's 3rd Armored Division had been on the go almost constantly, spearheading the U.S. VII Corps' drive across France and Belgium into the Third Reich.

Early the next morning, April 11, Colonel Castille again received another curious warning from division G–2: "Expect the unusual" in the Nordhausen region. Just past 8:00 A.M., Castille received an urgent message from the intelligence officer of Colonel John C. Welborn's task force: "Come quick! A big mainline railroad track leads right into a mountain. Lots of railroad cars loaded with 'stuff.' "

Now Castille was convinced that much of the 3rd Armored intelligence apparatus had taken leave of its senses. Why would a railroad track go into a mountain without emerging from the far side? And what was meant by the term "stuff"?

Castille hopped into his jeep and within a half-hour was at a place he learned was called Kohnstein Mountain. Along with Colonel Welborn and Lieutenant Colonel William B. Lovelady (a task force commander), Castille walked cautiously into the entrance of a tunnel, which immediately opened up into large chambers.

"It was a very elaborate factory layout," Castille recalled years later. "I recognized precision machine tools, complicated assembly lines, and the like—a massive, sophisticated, high-tech operation. It was obvious that we had indeed discovered something 'very unusual.' "[3]

Castille's "visit" was cut short by a terse message from Brigadier General Truman E. Boudinot, leader of the 3rd Armored's Combat Command B: "Report immediately to my advanced CP." His advanced CP (command post) consisted of a jeep, a driver, and a radio. On his way, Castille spotted several huge "sky rocket-looking objects" loaded on special railroad cars and two jet aircraft in a large hangar.

When Castille walked up to General Boudinot, who was sitting impatiently in his jeep, the intelligence officer began excitedly telling his combat command leader of the "fabulous finds" back at Kohnstein Mountain. Boudinot, whom 3rd Armored men regarded as a General George S. Patton clone, cut off Castille in midsentence: "Follow me!"

Boudinot raced off in a cloud of dust, and Castille leaped into his own jeep and sped after him. Minutes later, the vehicles halted at the gates of Caserne Boelcke, a collection of barracks outside Nordhausen. Each day at 4:00 A.M., hundreds of conscripted laborers who lived there had been marched to Kohnstein Mountain where they worked on the underground V–2 assembly lines.[4]

Camp Dora, a few miles north of Caserne Boelcke, also was overrun by elements of the 3rd Armored Division. Throngs of ecstatically happy inmates greeted the Americans.

In the meantime, Colonel Andrew Barr, the 3rd Armored Division's G–2, radioed word of the unique "find" in and around Kohnstein Mountain to higher headquarters, and the flash worked its way through the chain of command until it reached Colonel Holger Toftoy, chief of Ordnance Technical Intelligence, in Paris. Within hours, Toftoy formed Special Mission V–2. That group's function was to dash for the Harz Mountains on the heels of the 3rd Armored Division, grab the 100 long-range missiles reported to be in the underground factory, and ship them to the White Sands Proving Ground in New Mexico.

Ludy Toftoy was a man of action—even if he was not a rocket specialist. His expertise lay in submarine mines, and the army considered him their foremost specialist in the field. Toftoy and crews working under his direction had the perilous task of clearing harbors of underwater mines after D-Day in Normandy. At the risk of being blown into powder, the 41-year-old, bespectacled colonel had, in Cherbourg harbor, personally defused a type of German mine not previously known to Allied intelligence.

Leader of Special Mission V–2, who had the responsibility of "evacuating" the missiles, was 26-year-old Major James Hamill, a 1940 graduate of Fordham University with a degree in physics. Among those in his group were Major William Bromley, an ordnance officer who graduated from Stanford University in 1940 and who would be in charge of technical matters, and Louis Woodruff, an engineering professor at the Massachusetts Institute of Technology.

Major Hamill's Special Mission V–2 group headed for Germany and set up a coordinating base at Fulda, 58 miles west of the Mittelwerk in Kohnstein Mountain. Hamill had no inkling that he was embarking upon one of the most bizarre operations handed a U.S. Army officer in World War II.[5]

Also arriving in Nordhausen at the same time was Major Robert Staver, whom Colonel Gervais Trichel, the rocket chief in the Pentagon, had dispatched to Europe to locate and interrogate German V–2 scientists and engineers.

Staver promptly ran up against an obstacle. Although the Mittelwerk region had been cleared of the enemy by the 3rd Armored Division for two weeks, it still was classified as a combat zone, and there was an Army ban on technical investigators entering such a zone. This rule was designed largely to keep the Army from having a collection of dead American technical investigators on its hands.

Staver hurried back to Paris, where he took up his problem with Colonel Toftoy, a sort of father figure to the young major. Although Staver was not under Toftoy's command, Toftoy solved the problem by getting Staver assigned to an ordnance outfit that was already attached to the U.S. First Army in the Harz Mountains region.

On April 30, Staver left Paris, accompanied by Edward Hull, a General Electric Company engineer in Project Hermes. They reached the Harz Mountains, and in the days ahead investigated a large number of missile-related plants and facilities scattered about the region.

In the meantime, on April 24, Supreme Commander Dwight Eisenhower received an urgent order from the Joint Chiefs of Staff in Washington: "Preserve from destruction and take under your control records, plans, books, documents, papers, files, and scientific, industrial and other information and data belonging to German organizations engaged in military research." It was the first official confirmation from the Pentagon that the military was involved in a mad race with the Soviets and the British to grab German scientific know-how.[6]

A week later, on May 1, 1945, a solemn-voiced announcer broke into regular programming on Radio Berlin: "Our Fuehrer, Adolf Hitler, fighting to his last breath against Bolshevism, fell for Germany this afternoon in his operational headquarters in the Reich Chancellery.'' Actually, the fuehrer had died the previous day. With Soviet forces only a few blocks away, Hitler had put a Luger to his head and squeezed the trigger.

Although it had been obvious for weeks that the fuehrer's 1,000-year Reich was going to fall short of its goal by 988 years, the terse announcement of Hitler's death stunned Wernher von Braun, Walter Dornberger, and other missile experts at Oberjoch. They also were galvanized into surrendering to American troops; General Hans Kammler's SS men might yet shoot them or, equally grim, they might fall into the hands of the Soviet army.

Consequently, early in the morning of May 2, von Braun's brother Magnus, who spoke English fluently, climbed onto a bicycle and coasted down the long mountain road. Near the bottom of the incline, Magnus spotted a group of soldiers of the U.S. Seventh Army's 44th Infantry Division. Walking up to them, he asked in English: ''Pardon me. You are Americans?''

Private First Class Fred P. Schneiker of Sheboygan, Wisconsin, said they were indeed Americans.

''I represent a group of German rocket scientists, and we wish to surrender to the Americans.''

Schneiker eyed the stranger. Rocket scientists? What are rocket scientists? ''Well, I think you're crazy as hell,'' the American finally replied. ''But come along with me.''[7]

Magnus von Braun was escorted to a U.S. headquarters in the town of Reutte, where the young German was questioned by Charles L. Stewart, a counterintelligence agent. Years later, Stewart recalled:

> Magnus von Braun explained that his brother Wernher and some 150 of the top German rocket personnel were lodged in an inn behind German lines. They wished to join the Americans to continue their work in rocket development. They had selected the Americans, as they were favorably disposed to this country generally and also because this country was the one most able to provide the resources required for interplanetary travel.
>
> Furthermore, they were anxious to depart from the German side, as there was the possibility that an SS colonel in Innsbruck might eliminate them all, pursuant to last-minute Nazi orders to liquidate certain key German scientific talent to prevent them from falling into the hands of the Allies.[8]

At Oberjoch, seven of the Peenemünde Germans climbed into three passenger sedans at 2:00 on a chilly and overcast afternoon and, armed with safe-conduct passes given to Magnus by the Americans, they drove down the mountain and, ironically, through Adolf Hitler Pass. Those in the vehicles included Wernher von Braun and his brother Magnus; General Walter Dornberger and his chief of staff, Lieutenant Colonel Herbert Axster, an attorney in civilian life; Dieter Huzel and Bernhard Tessman, who had concealed the V–2 technical documents in the mine at Dörnten in the Harz Mountains; and Hans Lindenberg, an engineer.

A few miles along the way, the tiny convoy was met by two American military police jeeps and escorted to Reutte. There each man was assigned a comfortable room in a large requisitioned house. It was dark, and the Germans promptly went to bed.

Early in the morning, Wernher von Braun and the others were served their first American breakfast: fried eggs, bacon, toast, genuine butter, and real coffee. A meal fit for a king, von Braun reflected, after many months of food shortages in Germany.

Young GIs sat around the room, pecking at their own breakfasts and gawking wide-eyed at the youthful, charismatic German scientist who, they had learned, was the genius behind the development of the V–2. Von Braun joked and talked pleasantly with the Americans. One GI was overheard to say: ''He looks too fat and jolly to have launched anything more deadly than a kite.'' Like mischievous schoolboys the others snickered, and a second soldier ventured his view: ''If we haven't caught the biggest scientist in Germany, we've sure as hell bagged the biggest liar!''[9] Counterintelligence Agent Charles Stewart remembered:

> None of us had scientific backgrounds, but the magnitude of their [the scientists] discoveries and their potential for America's future was immediately apparent. We were dismayed when we could not arouse any interest in them at higher headquarters. Our first instructions were to the effect that they were to be thoroughly screened, a favorite solution of the military in dealing with high-ranking personnel if in doubt as to how to proceed.
>
> Our reply was to the effect that it made no difference if all were brothers of Adolf Hitler, because their unique knowledge made them extremely valuable from a national standpoint.[10]

In the meantime, some 550 miles to the north on May 5, Soviet troops led by Major Anatole Vavilov charged into Peenemünde and discovered that the great missile center was ''75 percent wreckage.'' Already battered by the heavy RAF bombing of August 1943, the test stands and laboratories had been partially blown up by elements of the Volkssturm, even as Red Army infantrymen were on the outskirts.[11]

Hard on the heels of Major Vavilov's force, Soviet technical intelligence teams rushed into Peenemünde, convinced that they would collar the Third Reich's foremost rocket specialists. To their dismay, they found that their quarries had flown the coop many weeks earlier.

The Soviet intelligence teams then began searching frantically for V–2 blueprints, reports, and other related documents and were frustrated to find only meaningless memos. They had no way of knowing that the intelligence bonanza they sought was hidden in a mine near Dörnten, in the American zone of advance.[12]

On May 7, 1945, two days after the Red Army overran Peenemünde, a delegation of Wehrmacht brass arrived at the headquarters of Supreme Commander Dwight Eisenhower in a red brick schoolhouse outside Reims, France. Within an hour,

General Alfred Jodl, Adolf Hitler's war-long military confidant, signed an instrument of surrender. At midnight the next day there ended, in Europe, a violent conflict that had raged since September 1, 1939.[13]

While Allied soldiers and most of the free world were rejoicing over the cessation of hostilities, several of the Peenemünde Germans were moved from Reutte to Garmisch-Partenkirchen, where they joined about 500 other rocket specialists whom the U.S. Counterintelligence Corps had rounded up from throughout the Alps. Garmisch, located at the foot of Germany's highest mountain, the Zugspritze, had been untouched by the war.

Wernher von Braun, Walter Dornberger, and the others were housed in relative comfort in a large building once used by the Wehrmacht as an administrative center. Armed American guards and the barbed-wire fence that surrounded the structure served as constant reminders to the Germans that they were not guests, but prisoners.

A day after the Third Reich capitulated, a bevy of American scientific investigators descended upon Garmisch-Partenkirchen after word was received that the Peenemünde scientists were being held there. One of the first to reach the scene was Richard W. Porter, leader of General Electric's civilian Project Hermes. For the past two weeks, Porter had been in Germany trying to track down Wernher von Braun, who headed Major Robert Staver's Black List of the most important German rocket scientists.

A native of Salina, Kansas, Porter was 32 years old and had earned his doctorate in electrical engineering at Yale University in 1937. General Electric executives who appointed him to the critical Hermes position considered Porter to be one of the nation's most brilliant young scientists.

Interrogators included the U.S. Army Air Force's Fritz Zwicky of the California Institute of Technology. Zwicky, born in Bulgaria and educated in Switzerland, was an astrophysicist who spoke German fluently.

Zwicky soon became disgusted with what he considered to be mass confusion. "There are too many technical teams, both American and British," Zwicky would later declare, "the members of which conducted interviews without any coordination." Von Braun, Dornberger, and the other German scientists "watched the unexpected and disorderly procedures of the British and American teams . . . and it became apparent that they considered our mission pretty much of a farce," Zwicky recalled.[14]

Much of the American interrogations focused on the Peenemünde men's political connections in the Third Reich. Von Braun explained that in his youth he had bowed to heavy pressure and reluctantly taken Nazi party membership, stressing that it was purely symbolic, that he had never played an active role. In the Third Reich, von Braun added, he would never have been permitted to conduct his rocket experiments had he not been on the rolls of the Nazi party.

Von Braun was questioned in a friendly manner. Civility was the password for the American scientific investigators. As Colonel Holger Toftoy had put it,

''We need their cooperation.'' For his part, von Braun was his customary congenial self—but in this poker game between victor and vanquished, he was playing his cards close to the vest.

Fully aware that the United States was at least 20 years behind in missile development, von Braun used that fact as his bargaining chip. He revealed just enough technical information to whet the appetite of the Americans. Significant details he kept to himself, a sort of carrot-and-stick technique.

It was clear to American investigators that von Braun, Dornberger, and other scientists were subtly seeking a proposal to conduct space research for Uncle Sam. ''We were interested in continuing our work in the United States, not just being squeezed like a lemon and then discarded,'' von Braun would declare later.[15]

Porter, Zwicky, and Toftoy in Paris did not have the authority to offer the rocket experts contracts for their services, a policy decision that would have to be made at the highest levels in Washington. That situation frustrated Toftoy. Unless the United States offered von Braun and other key rocketeers research contracts, they clearly were not about to provide the Americans with a wealth of scientific know-how that was in their heads—and presumably in documents that they had hidden. They were not prisoners of war and no charges had been filed against them, so as civilians they eventually could simply walk away and go home. Worse, they might sell their technical service to the Soviets.

While the Americans had as yet no clear directives and could make only vague promises to the Peenemünde contingent, the Russians, governed by the will of one man, Joseph Stalin, already had a plan of action. Each day, the Soviets broadcast over Radio Berlin that they were trying to locate Peenemünde experts and had highly desirable contracts waiting to be signed, that the rocket men would have exceptional living conditions and that their families would be able to join them in Russia.

Those interested in the Soviet proposal, the broadcasts declared, should apply to a Ludwig de Pinsky in Dresden. Partly due to the fact that the German missile scientists and engineers had never heard of de Pinsky, there was no mad stampede to sign up with the Soviets. Nor were the Peenemünde men enthused over the prospect of living and working in Russia.

A week after V-E Day, the missile experts at Garmisch had dwindled from about 500 to 200. Some of those departing had their addresses and those of close relatives recorded in the event their services were sought later; they were then furnished transportation to their homes. Others simply climbed over the fence at night and melted into the population; no real effort had been made to halt the ''escapes.''

Only the foremost rocket experts remained at Garmisch. Now that they had been collected, no one seemed to know what to do with them.

Wernher von Braun (inset) at age 16 already was a rocket genius. (U.S. Space and Rocket Center)

Robert H. Goddard, a pioneer American rocket innovator, at work in his New Mexico laboratory in 1935. Goddard's efforts were largely ignored by the U.S. government and armed forces. (National Archives)

Deutsches Reich

Flugzeugführerschein

Nr. 663

für Freiherr von Braun, Wernher

(Stempel der ausstellenden Behörde)

(Stempel der ausstellenden Behörde)

Wernher Freiherr v. Braun
(Eigenhändige Unterschrift des Inhabers)

Charlottenburg, den 20. 9. 1933.
(Ort)

(Stempel der ausstellenden Behörde)

Der Oberpräsident
Im Auftrage:

(Unterschrift)

Wernher von Braun's German civilian pilot's permit. (NASA)

Thirty-year-old British scientist Reginald V. Jones (left) warned that Germany had long-range rockets. (Courtesy Professor R. V. Jones)

Winston Churchill's chief scientific advisor Lord Cherwell (right) doubted that Germany had V-2 rockets. (National Archives)

This aerial photo of Peenemünde, taken on June 23, 1943, was the first to reveal German long-range rockets (Figure A). Figure B points to where rockets were stored. (U.S. Air Force)

Entrance to one of two main tunnels at the Mittelwerk, a huge underground plant in the Harz Mountains where V-1s and V-2s were assembled, free from air reconnaissance and bombing. (U.S. Army)

V-2 rocket assembly line in the Harz Mountains' underground factory. (U.S. Army)

U.S. 3rd Armored Division Lieutenant Colonel William A. Castille was the first American to enter V-2 underground plant. (Courtesy W. A. Castille)

Colonel John C. Welborn (right), 3rd Armored Division, was with Castille when they discovered the Harz Mountains' V-2 factory. (Courtesy Elizabeth Welborn)

General Electric's Richard W. Porter played a key role in rounding up German rocket scientists. (Courtesy R. W. Porter)

German Armaments Minister Albert Speer (left) saved Wernher von Braun and two rocket associates from execution by the Gestapo. (National Archives)

Wernher von Braun (right), with his arm and shoulder in a cast, and his brother, Magnus, surrender to Private First Class Frederick P. Schneiker of the U.S. 44th Infantry Division in May 1945. (U.S. Army Missile Command)

Harry S Truman being sworn into office after President Roosevelt's death in April 1945. Truman gave approval for Operation Overcast. (U.S. Army)

Army Chief of Staff George C. Marshall (right) authorized bringing Peenemünde rocket experts to the United States. (U.S. Army)

Major James Hamill (left), a key figure in Operation Overcast, and Wernher von Braun after the Peenemünde team arrived in the United States. (U.S. Army)

Colonel John C. Nickerson, Jr. (left), jeopardized his career by challenging an adverse missile decision by Secretary of Defense Charles Wilson. (U.S. Army Missile Command)

Major General Holger N. "Ludy" Toftoy (right) was in charge of the search for German rocket experts at the close of war in Europe. (U.S. Army Missile Command)

Leading figures on the Peenemünde rocket team at Huntsville. From the left: H. H. Maus, Kurt Debus, Eberhard Rees, Hans Hueter, William Mrazek, Werner von Braun, Walter Haeussermann, Erich Neubert, Ernst Geissler, Karl Heimburg, Helmut Hoelzer, and Ernst Stühlinger. (U.S. Army Missile Command)

Peenemünde team rapidly became civic leaders in Huntsville. Walter Wiesman (left) was president of the Huntsville Junior Chamber of Commerce in 1953, two years before he became a U.S. citizen. (Courtesy Walter Wiesman)

President Dwight D. Eisenhower presents the Distinguished Federal Civilian Service award to Wernher von Braun in 1959. Secretary of the Army Wilber M. Brucker looks on. (U.S. Army)

First seven American astronauts. Front, left to right: Virgil "Gus" Grissom, M. Scott Carpenter, Donald K. Slayton, and L. Gordon Cooper. Back, from the left: Alan B. Shepard, Walter M. Schirra, and John H. Glenn. (NASA)

Speaking to Congress on May 25, 1961, President John F. Kennedy proposes landing a man on the moon within the decade. Behind Kennedy are Vice President Lyndon B. Johnson (left) and Speaker of the House Sam Rayburn (right). (NASA)

Vannevar Bush (left) called Kennedy's moon-landing proposal "a stunt." (National Archives)

President Kennedy touring the Marshall Space Flight Center in September 1962. Wernher von Braun (left), Major General Francis J. McMorrow, and Vice President Lyndon Johnson accompany him. (NASA).

Moon-landing team (from left): Neil Armstrong, Michael Collins, and Edwin ''Buzz'' Aldrin. (NASA)

Booster stage of Saturn 5 rocket dwarfs two men (lower left). (NASA)

Florida's John F. Kennedy Space Center in May 1966. The Saturn 5 had just been removed from the 54-story high Vehicle Assembly Building to be inched upright three and a half miles to the launchpad. (NASA)

Jubilation reigned in Houston's Mission Control when *Apollo 11* splashed down in the Pacific. From right to left: Robert Gilruth, George Low, Chris Kraft, and Samuel Phillips. (NASA)

# 12

## "They're 25 Years Ahead of Us!"

On February 3, 1945, the Allies' "Big Three"—President Franklin Roosevelt, Prime Minister Winston Churchill, and Premier Joseph Stalin—held a conference at Yalta, a famous Black Sea resort, during which they agreed on a plan to divide Germany into four unequal portions.

The United States, Great Britain, Russia, and France each would occupy a part, with Berlin (inside the Soviet zone) to be jointly administered by the four nations. Since American armies had driven far into what would be the Russian occupation zone, the U.S. First Army soon would have to pull back and relinquish a huge chunk of territory (including the Harz Mountains) to the Soviets.

Now, three months later, chaos gripped the Third Reich. Scores of once majestic cities were wastelands of fallen masonry. Hundreds of towns and villages had been pounded to splinters and pebbles by Allied air and artillery. Bridges were destroyed. Railroads were shambles. Millions of ragged, dazed German civilians stumbled along the highways and byways. Most of the meandering bands which had fled to escape the fighting now were headed for uncertain futures in what had been their homes. Seldom, if ever, in history had a land and a nation been so utterly devastated.

Within this panorama of destruction and mass confusion, U.S. Army Major Robert Staver, for nearly a month, had been scouring the Harz Mountains in search of the V–2 scientists and engineers on his Black List. Scores of them had not gone to the Bavarian Alps with Wernher von Braun; they merely had melted into the villages in the vicinity of Bleicherode. Staver could not uncover a single one of his targets.

In the meantime, Major James Hamill, who was in charge of Special Mission V–2, was confronted with an equally frustrating situation in his efforts to "evacuate" 100 huge missiles. His was a delicate task—in many ways. Hamill kept recalling the words of Colonel Holger Toftoy: "High-level agreement among the Allies is that nothing is to be moved out of the Russian zone. However, unofficially, I'm telling you to see that those 100 V–2's get to Antwerp. Get all the materials you can without making it too obvious that we've looted the place.''[1]

Colonel Toftoy was a dedicated and flexible man who lived in the real world. As with most high-ranking American officers, he knew that the mighty struggle for postwar dominance between the free nations of the Western world and Russia and its Communist allies already had begun. Even while hostilities were still raging, Russia flouted the Allied declaration for a free Europe and installed Communist-dominated governments in Poland and Romania. At the V–2 test-firing center at Blizna, the Soviets demonstrated that they had no intention of cooperating with Great Britain and the United States. So to Colonel Toftoy, 100 long-range missiles, along with their technical documents, falling into the hands of the Russians was an unthinkable specter.

While reconnoitering the Mittelwerk after it had been overrun by the U.S. 3rd Armored Division, Major Hamill, Major William Bromley, and Louis Woodruff promptly discovered that their task of removing the V–2s before the Soviets arrived was mind-boggling in scope. No date had been announced for the American pullback from the Russian zone, but Hamill, Bromley, and Woodruff believed it to be June 1.

One major obstacle was that 100 V–2s were not sitting conveniently in the tunnels ready for shipment to Antwerp. The Special Mission V–2 team would have to select components found in Mittelwerk for later assembly of the long-range missiles. This would be an impossible task without the parts list and other V–2 technical documents that still had not been uncovered by the American technical intelligence teams. Nor had a single key German rocket specialist who might provide guidance in the selection of parts been located in the Harz Mountain region.

There was a major railhead at Nordhausen that had escaped destruction by Allied bombers, but not nearly enough trucks could be rounded up for transporting the tons of V–2 parts from the underground factory to the railhead. Everything had to be done on a hush-hush basis, for ''liberating'' (as the GIs termed such actions) the V–2s was a covert operation not officially sanctioned by the Pentagon.[2]

On reconnaissance missions, Hamill, Bromley, and Woodruff discovered more depressing news: most of the main railroad bridges and large portions of track leading from Nordhausen to Antwerp were twisted wreckages, mute testimony to the accuracy and ferocity of Allied bombers. It was estimated that 340 railroad cars would be needed to ship the missile components to Antwerp. Sixteen Liberty ships then would be required to haul the Mittelwerk loot across the Atlantic.[3]

Undaunted, Hamill searched for a solution to the transportation problem. The bombing of all but one of the key railroad bridges on both sides of Nordhausen turned into a ray of hope: hundreds of German railroad cars were marooned in the region. But Hamill's hope was shattered. The Army Transportation Corps planned to move this rolling stock, most of it in passable condition, out of the region and westward into the American occupation zone, using the lone intact railroad bridge.

In the meantime, the Mittelwerk had been, in essence, "open to the public." Furious inmates from the Dora and Nordhausen concentration camps who had been working on the V–2 assembly lines returned to wreak vengeance by destroying thousands of irreplaceable missile components and machine tools. Civilians in the battered town of Nordhausen trekked to the underground factory in droves and stole light bulbs, missile components, and, in the words of an American officer, "about anything else that wasn't nailed down." Even Soviet intelligence agents, concocting reasons for being behind American lines, traipsed in and out of the underground factory.

Two weeks after the Mittelwerk was captured, Major William Bromley secured the services of a company of infantry from the U.S. 5th Armored Division and a tight cordon was thrown around the premises. No one could enter without a pass from the Special Mission V–2 leaders. The GIs had been told only that "important technical items" were in the tunnels.

Now that he was not being disturbed by a constant flow of people tramping in, around, and out of the factory, Louis Woodruff, the electrical engineering expert, was able to isolate many of the needed missile components based on his intense study of U.S. intelligence documents. However, the control systems that guided the V–2s were manufactured elsewhere in the region, and without these crucial items it would be impossible to reconstruct a V–2 in the United States.

Hasty probing revealed that just before the American armored spearheads had charged into the Harz Mountains, the control systems already manufactured had been hidden around the area. Hamill and Bromley rapidly organized reconnaissance parties which fanned out into the region. They brought back scores of the critical control systems which, in keeping with the fuehrer's orders, had been concealed in unlikely places—barns, houses, garages, even a *bierstube* (beer hall).

Now the Special Mission V–2 team was ready to load onto railroad flatcars the many thousands of missile components and subassemblies, but this work would require perhaps, 200 men with a knowledge of machinery. That problem was resolved when Major Bromley obtained the services of the 144th Motor Vehicle Assembly Company which was then based at Cherbourg, France, near where the American amphibious and airborne assault forces had landed on D-Day. Cherbourg was some 760 miles from Nordhausen, but the 144th unit was ordered to get on the road immediately. There was not a minute to be lost. Only 12 days remained before the Red Army was to take over the region.

There were more maddening obstacles and delays. The Kohnstein Mountain tunnel through which the railroad tracks ran had been clogged with earth, rock, and debris for a mile, and this had to be cleared away before loading and packing operations could begin. Nearly 200 laborers were hired at the Dora and Nordhausen camps and paid to do the heavy work.

When the truck outfit from Cherbourg arrived, its members joined with the 319th Ordnance Battalion to sort out components and sections of the V–2, a job they were adapted to because of their machinery backgrounds. Altogether, more than 1,000 men were involved in the feverish underground project in what one American officer called ''organized chaos.'' Bromley and Hamill had hoped to use around-the-clock shifts, but the tunnel ventilation system broke down, so work was limited to eight-hour days, much to the annoyance of the Special Mission V–2 team which was engaged in a desperate duel with the clock.

Late in the afternoon of May 19, Hamill and Bromley received staggering news. Early the next day, as planned, the Army Transportation Corps would begin moving the hundreds of usable railroad cars in the region across the lone surviving bridge outside Nordhausen and into the American zone. It appeared that all of the thousands of tedious manhours, backbreaking work, and improvised tactics would go for naught, and the V–2 components would be taken over by the Russians.

Hamill, Bromley, Woodruff, and other key members of Special Mission V–2 had heavy hearts that night. Then a curious event occurred: during the hours of darkness, the railroad bridge was blown up with dynamite. After dawn, the Transportation Corps trainmen discovered that they could not move the rolling stock.

The Transportation Corps officers in Nordhausen were furious. Hamill and Bromley swore that they had no idea who could have committed the blasting but ventured to guess that it might have been the work of the German underground. But how could the underground, if indeed there were one, know that the rolling stock was to be moved over that bridge after dawn? Hamill and Bromley merely shrugged.

Now the Special Mission V–2 leaders were provided with time to consult with the Transportation Corps. By now, Jim Hamill and Bill Bromley had been provided with a card from Supreme Headquarters, Allied Expeditionary Force (SHAEF), authorizing them to collect and ship German equipment. The signature (or facsimile) of General of the Army Dwight D. Eisenhower leaped out at the Transportation Corps officers, and they had an immediate change of heart about the railroad rolling stock in the Nordhausen region. They would be most happy to relinquish rights to the cars and turn them over to Hamill and Bromley.

Within hours, the 1186th Combat Engineer Company repaired the blown railroad span and, working at feverish pitch, built a bridge that connected the underground factory tracks with the main railhead at Nordhausen. In order to operate the trains, Hamill and Bromley recruited former German railroad employees, who were only too happy to find a paying job. They proved to be most cooperative and efficient.

Early in the morning of May 25, 1945—five days before the Soviets were expected to arrive—the first train, loaded with V–2 components, chugged out of Nordhausen. It rolled to Erfurt, where men of the U.S. Railway Service took it over and steered it on to the port of Antwerp.[4]

During the next few days, nine other trains, each averaging 34 cars, steamed out of Nordhausen for the trip to Antwerp. The last train departed on the evening of May 31—six hours before the Russians reportedly would arrive. Altogether, there were 341 rail cars carrying a cargo of nearly 500 tons, the largest single movement of captured German equipment during and after World War II.[5]

It was still too early for the Special Mission V–2 team to rejoice over its coup. More trouble lay ahead as the result of an earlier agreement between the United States and England stating that, whichever nation uncovered V–2s, the other would get half of them.

In Paris, Colonel Holger Toftoy, aware of the prior agreement with the Brits, nevertheless had not notified his English counterpart of the ''evacuation'' of the long-range missiles from the Mittelwerk. And he recommended to his superiors in secret communications that the entire shipment of missile parts be sent to the United States.

It was Toftoy's reasoning that England, whose treasury was seriously depleted in the wake of nearly six years of all-out war, would not be in a position to exploit rocket research and development, and that the Western Allies would be better served by shipping all of the V–2 components to the United States.

Now two British undercover agents spotted the V–2 shipment on the Antwerp docks waiting to be loaded on ships, and they promptly notified their headquarters in London. High-ranking British officers at General Dwight Eisenhower's SHAEF headquarters vigorously protested the American action, but before anything could be done to halt the shipment, the crates were on board the Liberty ships and heading for the port of New Orleans in the United States.

Meanwhile, Major Robert Staver, on the night of May 12, 1945, scored his first success in his intense search for German missile experts in the Harz Mountains. Acting on a tip, he located Karl Otto Fleischer, general manager of the Elektromechanische Werke, living in Nordhausen.

Fleischer agreed to cooperate with the Americans, but he kept to himself the fact that he was the only person in the Harz Mountains who knew that the 14 tons of V–2 research documents were hidden in the Dörnten mine. Dieter Huzel and Bernhard Tessman, who had hidden the papers, had confided the secret to him before departing for the Bavarian Alps.

Fleischer took Staver and Edward Hull of Project Hermes to Eberhard Rees, who had been a department chief in charge of the Peenemünde plant. Rees, too, was cooperative and informed Staver and Hull that Klaus Riedel, former chief of the rocket engine and structural design department at Peenemünde, was in jail at Saalfeld, 55 miles from Nordhausen.

Major Staver rushed to the Saalfeld jail and found that Riedel had been a victim of mistaken identity. U.S. counterintelligence agents thought that he had been involved in the development of a ''bacteria bomb.'' Forty-eight hours later, on May 16, Staver secured Riedel's release through the intervention of the U.S. Military Government.

Back at his home in Nordhausen, 34-year-old Klaus Riedel spelled out for

Staver and Hull a lengthy rundown on German rocket development. Like Wernher von Braun in Bavaria, Riedel declared that the Peenemünde group's goal really had not been weapons but rather "passenger-carrying rockets, trips to space stations, journeys around the moon, and daring explorations of outer space."[6]

During the prolonged conversations, the husky, blond Riedel suggested that it would be a prudent and beneficial move by the U.S. government if it shipped a contingent of the Peenemünde missile experts to America to resume their research and development projects. Otherwise, Riedel pointed out, the Soviets might round up the rocketeers, and take them to Russia. That made a great deal of sense to Major Staver, who knew that the Red Army soon would arrive to take over the Harz Mountains region.

Acting on Riedel's suggestion, Major Staver fired off a cable to Colonel Gervais Trichel, his boss in the Pentagon, urgently recommending that 100 of the Peenemünde scientists and engineers be offered short-term contracts and imported to the United States. "The thinking of scientific directors of this group is 25 years ahead of ours," Staver said. "They have begun development of A-10 to have thrust of about 220,000 pounds," he continued. "Later version of this rocket should permit launching from Europe to U.S." On a more immediate priority, the German experts could be put to work refining the revolutionary Wasserfall antiaircraft rocket for use in the Pacific War. What was more, Staver said in his cable, such a recruitment program would prevent the rocket experts from joining up with "other interested parties"—meaning the Soviet Union and, to a lesser degree, Great Britain.[7]

Colonel Trichel had been thinking along the same lines in recent weeks, so he passed Staver's proposal along the chain of command with his enthusiastic endorsement. Most army brass, including Chief of Staff George C. Marshall, looked favorably upon the recruitment project, for the invasion of the Japanese home islands was expected to be bloody and savage and could last for three or four more years. In the Pacific, Supreme Commander Douglas MacArthur predicted one million additional American casualties before Japan was conquered.[8]

Subtly, the Pentagon began smoothing the way for importing the German scientists. On May 22, Major General Clayton Bissell, the Army assistant chief of staff for military intelligence, dispatched two colonels to call on Assistant Secretary of State Frederick Lyon. They diplomatically informed Lyon that the Army intended to bring German rocket experts to the United States and, since they would technically be under arrest, State Department visas would not be necessary.[9]

American intelligence, Lyon was told, had uncovered evidence that German rocket blueprints and a mass of other technical materials already had been passed along to the Japanese high command, so it was crucial that the Peenemünde experts be brought to the United States in an effort to beat the Japanese in rocket development. Secretary Lyon wrote to Secretary of State Cordell Hull: "One can readily appreciate the necessity of having these men [Germans] work here for a temporary period."[10]

Bringing the German scientists to the United States presented no major problem to the Justice and State departments, but the Pentagon veiled the project in the deepest secrecy, fearful of a public outcry against "bringing Nazis into the country." There was a second reason for the top-secret status: to keep the Soviets in the dark.

On May 28, Undersecretary of War Robert Patterson, a former federal judge, approved Colonel Trichel's plan, as recommended by Major Staver. "Every possible aid should be given to prosecuting the war against Japan," Patterson wrote. However, he cautioned: "These men are enemies and it should be assumed that they are capable of sabotaging the war effort. It is assumed that these scientists will be under strict surveillance while here."[11]

In the meantime back in Germany, Major Bob Staver was continuing his desperate efforts to uncover clues that would lead him to where the crucial V–2 documents had been secretly stashed. He was aware of Major James Hamill's Special Mission V–2 (to ship long-range missiles to New Mexico) and knew that, unless the blueprints and related data were located, it would be nearly impossible to reassemble the V–2 parts and test-fire the missiles.

As the days rolled past, Staver continued to interrogate Karl Otto Fleischer, who helped to locate many items of rocket equipment but remained silent on the burial site of the priceless documents. Staver suspicioned that Fleischer and Rees knew where the V–2 materials had been stashed and decided to try to hoodwink them into revealing the location. Since there were no civilian telephone lines or other communications between Nordhausen and the Bavarian Alps, Fleischer and Rees were unaware that Wernher von Braun, General Walter Dornberger, and other missile experts were in the custody of the U.S. Army.

Now Staver pulled a notebook from his pocket and began reading what seemed to be a report from U.S. counterintelligence agents but actually was a report created by Major Staver himself: "Von Braun, Steinhoff, and all others who fled to the south have been interned at Garmisch. Our intelligence officers have talked to Colonel von Ploetz [General Hans Kammler's aide], General Dornberger, and General Kammler. They told us that many drawings and important documents were buried underground in a mine somewhere [around the Harz Mountains]."[12]

Staver studied the two Germans. Riedel, who knew nothing about the secret cache, remained expressionless. Fleischer flinched like a raccoon caught at night in a flashlight beam but remained silent. The major was now convinced that Fleischer knew the burial site.

Twenty-four hours later, Staver got word that Fleischer wanted to see him and that the German was at the home of a Catholic priest in a nearby village. Staver leaped into his jeep, suspecting that his target was about to crack, and raced to the house.

Fleischer, looking drawn and tired, came down from upstairs, escorted Staver outside, and admitted that he knew where the priceless V–2 documents were concealed. Qualms that he would be a traitor to Germany had kept him from disclosing this knowledge earlier, he explained. But now that the generals had told the Allies of the cache (which they had not done), he would cooperate. Fleischer admitted, however, that he did not know the precise location of the most important missile documents in the world.[13]

Undaunted, Major Staver got hold of a civilian automobile, filled it with gasoline, provided Fleischer and Eberhard Rees with passes that would allow them to remain outside after the 8:00 P.M. curfew for civilians, and told them to drive around the Harz Mountains to locate the mine. Rees suggested that Staver remain behind, for the presence of an American officer might inhibit German civilians from giving information that might lead to the burial site.

Bob Staver agreed and stood by as Fleischer and Rees drove away. But the major was deeply worried. Would the American ever see the two scientists again? What would keep them from using the car, gasoline, and passes simply to drive to other parts of Germany and take refuge in the chaos of the shifting civilian population? Or worse, would Rees and Fleischer hot-foot it to the Soviet zone and offer the Russians their services—and the precious documents?

Fleischer and Rees were embarked on something of a needle-in-a-haystack search. The Harz Mountains were dotted with hundreds of mines, many of their entrances concealed with vegetation. Civilians were of minimal help. After crisscrossing the Dörnten region for almost the entire day, the two Germans finally located the elderly caretaker of one mine who insisted repeatedly that, although his excavation was blocked by tons of rock, there was no cache hidden in it.

A heated argument raged for nearly an hour. Fleischer declared (falsely) that he had been in charge of the entire German rocket program and that high-ranking German generals had sent him to the mine with orders to retrieve the documents inside and hide them at another place. The caretaker mulled over Fleischer's pronouncement; then, with an "*Ach*" and a deep sigh, he relented. Yes, three trucks had brought hundreds of boxes to the mine six weeks earlier. The tunnel inside had been sealed by blasting, and the caretaker himself had a day earlier dynamited a few tons more of rock onto the entrance.

Karl Fleischer and Eberhard Rees hopped into their car and raced back to Nordhausen, where they told an excited Major Staver of their find. The major rushed to the mine site (about 30 miles from Nordhausen), rounded up a large crew of unemployed miners, who were assured that they would be well paid, and organized a work force into three shifts. Then he secured the services of a platoon from the nearby 83rd Infantry Division to mount an around-the-clock guard at the site. Orders were given to shoot any unauthorized person who tried to make off with any of the boxes in the mine.

Time was of the essence. In accordance with the Yalta Agreements, the British were to take over the Dörnten region in only six days, on May 27. If he failed to haul off the 14-ton cache, the V–2 documents would fall to the British by default.

Burrowing through the hundreds of tons of rock and dirt that clogged the tunnel would be a prodigious task. It was so narrow that only two or three men could work at one time. They would shovel feverishly until nearly exhausted; then another two- or three-man team would relieve them.

With the digging well under way, Bob Staver caught a ride in an airplane to Paris, where he excitedly told Colonel Joel Holmes, chief of the Ordnance Technical Division, of his discovery. Within two hours, Holmes arranged to have two large semitrucks to go immediately to the Dörnten mine.

Staver hitched another airplane flight back to Nordhausen, then jeeped to the mine site where the digging was proceeding at a frantic pace. There was only one day left before the British took over.

Suddenly, there were loud shouts from within the tunnel: the diggers had broken through to the chamber where the hundreds of boxes had been hidden. Staver rapidly organized the miners into a caravan of back-packers, and they carried out the boxes and stacked them in front of the mine entrance. Now Staver was suddenly alarmed: the two semitrailers that were supposed to have been there four hours earlier had not arrived.

Frantically, the major telephoned the U.S. Ninth Army ordnance officer, who rushed six GI trucks to the mine site. The loading of the boxes continued through the night. At 6:15 A.M., less than four hours before the Brits were to take over, Staver led the six-truck convoy along the narrow, winding roads of the Harz Mountains to Nordhausen. On the way, it passed British vehicle convoys that were heading in the opposite direction to occupy the region.[14]

In the meantime, on May 19, 1945, Herbert Wagner and four associates who had been with the Henschel Aircraft Company in Germany, arrived in Washington, D.C., under a cloak of secrecy. They were the vanguard of hundreds of German technical experts to be imported to the United States, even though no formal government policy had been adopted.

Wagner had been interrogated and ''recruited'' by Admiral Ernest King's Navy Technical Mission in Europe. A report was sent to the Pentagon explaining that Wagner's knowledge of the Hs 2T3 (a radio-controlled air-to-air missile) could be fully exploited only if he were brought to the United States. Wagner's know-how and skills, the report declared, were ''unmatched anywhere in the world.''

On arrival, Wagner was hustled to a Washington hotel suite, where Admiral King's experts grilled him intensely for four weeks. Then Wagner was taken to a secluded estate on Long Island to work on a Navy top-secret missile project.

# 13

# Smuggling a Trainload of Missile Experts

Major Robert Staver hardly had time to bask in the accolades that were being heaped upon him for locating and carting off the mother lode of V–2 documents in the Dörnten mine than he received another daunting mission from Colonel Holger Toftoy in Paris. On May 27, 1945, Toftoy instructed the young major to evacuate all German missile technicians and their families from Thuringia, the region in Germany that covers 4,540 square miles, including the Harz Mountains, and take them to the American Zone.

It would seem to be an almost impossible task. There were nearly 3 million people living in Thuringia, and even finding the technicians would be tedious and time consuming. As with the Dörnten mine caper, time was crucial. This valuable human reservoir of technical know-how had to be gotten out of the zone that was to be occupied by the Russians in four days.

Another difficulty confronting Staver was that he had received no authorization to offer permanent employment to the technicians, and there was doubt about whether the Germans would want to be uprooted on a moment's notice and spirited away to an unknown destination and an uncertain future.

Convinced that his task was akin to trying to swim up Niagara Falls, Major Staver nevertheless plunged ahead with customary zeal. Then Lady Luck smiled on the Americans. Due to high-level wrangling among the Allies, the Soviet move into Thuringia would be postponed for three weeks, until June 21. Now Staver had some breathing room—but not much.

To speed up the search, Staver brought Ernst Steinhoff and a few other German scientists from Garmisch to Nordhausen. There, working with Eberhard Rees and Klaus Riedel, they drew up a list of some 500 names of technicians living in Thuringia who would be helpful to U.S. missile research and development. Armed with this list, Steinhoff, Rees, and Riedel, each with a driver and a German who knew the men to be contacted, fanned out over Thuringia.

Most of those contacted balked. What did the United States have to offer them in the future? Since no official American policy had been handed down from on high, the only reply was that the technicians and their families would be whisked out of Thuringia before the Soviets arrived, along with a promise that they would be well fed and housed in the American zone.

Major Staver and his men obtained few positive results. So Wernher von Braun was flown to Nordhausen to lend his prestige and powers of persuasion to the large-scale recruitment project. His appeal proved to be magic, for technicians signed up in droves.

In the meantime, Richard Porter, the young head of General Electric's Project Hermes, flew in from Garmisch to assist in the Thuringia roundup. When hundreds of technicians had cast their lot with the Americans, another daunting problem confronted Staver and Porter: transporting the technicians and their families to the Nordhausen railroad station to board a train for the American zone. By begging and pleading at one Army headquarters after another, Porter and Staver managed to round up a varied assortment of some 325 GI vehicles.

Now time was running out. The Soviets would take over Thuringia in only three days. A soldier driver was assigned to each of the fleet of vehicles and an around-the-clock taxi service was put into operation. Dashing about the region, the taxis would screech to a halt before the targeted technician's home and he and his wife and children would be given 15 minutes to grab what personal items they could carry.

Some of the targeted Germans had to get to Nordhausen on their own. One scientist, Ernst Stühlinger, responded to an insistent banging on his front door and was faced by an American captain who wasted no time on small talk. "You are to leave immediately," the captain declared. When Stühlinger protested that he had no means to get his family to Nordhausen, the American replied, "Then take the mayor's car. Within 36 hours, the Russians are coming and he won't need it!"

Back at Nordhausen, 12 hours before the mass exodus of German technicians was to begin, Hamill, Staver, and Porter debated the wisdom of blowing up the underground V–2 factory to deny it to the Russians. The Allies had agreed earlier that all industrial plants were to be left intact by withdrawing forces, but how could the blame be pinned on anyone if a mystery explosion in the middle of the night left Mittelwerk a mass of twisted wreckage and destroyed the thousands of V–2 components remaining in the tunnels? Reluctantly, the demolition scheme was abandoned—and the future course of history may have been unwittingly altered.

At noon on June 20, 1945, the missile experts and their families were gathered at the Nordhausen railroad station. A veil of thick apprehension hovered over the crowd. Stories of atrocities committed by the Red Army in Berlin and elsewhere had seeped through to Nordhausen, and the Soviets were due to arrive in less than 24 hours.

A long line of railroad passenger cars stood waiting along the tracks, but the locomotive failed to appear. The clock ticked on: one hour, two hours . . .

three . . . four. Where was the locomotive? Suddenly, a loud cheer erupted from the crowd—the engine was chugging into Nordhausen.

Just past 6:00 P.M., less than five hours before the Russians were to occupy the region, the train, crammed with missile specialists and their families, crossed the Werra River over a repaired bridge and sped on to Witzenhausen, a small town 40 miles southwest of Nordhausen in the American zone.

Through prodigious endeavor and ingenious thinking, the Americans now seemed to be holding all the aces in the high-stakes missile-secrets poker game with the Soviets. Components for 100 V–2s had reached New Orleans and were being loaded on trains for shipment to White Sands in New Mexico. The tons of V–2 documents (including 65,000 drawings for the development of the original V–2) from the Dörnten mine were in the hands of American experts at the Aberdeen Proving Ground in Maryland. Wernher von Braun, the world's leading rocket expert, and his principal associates were under the U.S. Army's wing in Garmisch, and now most of the other Peenemünde specialists had been smuggled out of Thuringia and into the American zone.

When Georgi Malenkov's Soviet technical intelligence teams hurried into the Harz Mountains on the heels of the withdrawing Americans, they were elated to find that the underground V–2 factory was in nearly perfect working order. Thousands of machine tools for constructing missiles were intact and operable, although the Soviets knew that the Americans had removed a large number of components.

In rapid fashion, the Russians located German rocket technicians who had chosen to remain in Thuringia and, with the offer of good wages, enticed them to go to work assembling long-range missiles in the maze of tunnels and in requisitioned factories in the region. In the chaos of postwar Germany, scientists, engineers, and technicians had their choice of going to work for whomever could pay them or, along with their families, starving.

All the while, cooperation between the Soviet Union, and the Western Allies—a bizarre marriage of convenience that had resulted from mutual goals in the war against the Third Reich—was eroding. Since the conclusion of hostilities in Europe, the four major Allies had paid lip service to sharing German scientific secrets and equipment. Scores of Russian technical intelligence teams roamed freely in the American, British, and French zones, but when American teams visited facilities in Soviet-occupied Germany, their requests to inspect important scientific installations and factories were stonewalled.

Although distrustful of the Soviets, Washington continued to cooperate with its wartime Communist ally. So when the Kremlin asked for the Americans' Black List (a compilation of 15,000 German scientists and engineers), a decision was reached to hand over a copy to the Soviets (although some of the most important names were deleted).

In London, Prime Minister Winston Churchill was stunned by what he considered to be an incredibly naive approach by Washington in giving the Russians the Black List. A lifelong crusader against communism, the British Bulldog nevertheless had clasped the Soviet Union to his bosom back in 1941, when Great Britain was a lion at bay and on the verge of being crushed by Hitler's seemingly invincible Wehrmacht. Now the war was over. So on June 26, 1945, Churchill cabled President Harry Truman that Britain would share scientific intelligence only with the United States. Truman, confronted with a mass of intelligence on Joseph Stalin's double-dealing, agreed.

As American, British, French, and Soviet technical experts sifted through the V–2 documents, studied captured equipment, inspected German factories, and interrogated German scientists and engineers, it became awesomely clear that the past had been merely a prelude to an unimaginable future for rockets. One of the more fantastic ideas of German scientists had been a winged rocket that Eugen Sänger and Irene Bredt had first proposed in August 1944, at a time when American spearheads were racing across France toward the borders of the Third Reich. The top-secret report proposed an antipodal bomber, a revolutionary concept. It envisioned a 92-foot-long, 220,000-pound craft to be launched from a sled driven by rockets developing 1.345 million pounds of thrust. The sled would send the bomber into the air at 1,000 miles per hour; then its own 200,000-pound-thrust liquid oxygen–gasoline oil rocket engine would boost it to a speed of 13,700 miles per hour and an altitude of over 160 miles.

If built, the manned antipodal bomber would skip along the top of the atmosphere (where no human had ever been) like a stone on a pond, reaching New York with a bomb load of six tons. Sänger and Bredt calculated that the entire flight, from takeoff to a return landing in Germany, would require only 80 minutes.[1]

After Sänger and Bredt completed their work, 100 copies of the 409-page report were made, and an authorized functionary slapped a big rubber stamp on the front page:

> This is a State secret and must be passed personally hand to hand. . . . Any copying, photographing, etc. verboten. To be kept in locked steel safe in rooms which are guarded 24 hours a day. Any neglect of these regulations will be most severely punished.[2]

A list of 80 recipients was drawn up and the state secret was on its way by couriers in double envelopes with return receipt. Among the German VIPs receiving the Sänger-Bredt report were Wernher von Braun and Walter Dornberger at Peenemünde; Willy Messerschmitt, designer of the Third Reich's frontline fighter planes; Kurt Tank of Focke-Wulf aircraft; Walther Dornier of Dornier aircraft; and Julius Mader of Junkers aircraft.

The list read like a Who's Who in German missile development and aircraft design, but in the beleaguered Reich, with enemy armies closing in from three sides, few, if any, had the time or inclination to wade through the 409 pages of technical detail.

If the German scientific and technical community was not overly enthused about the concept of the antipodal bomber, after Victory in Europe Day, the Soviets certainly were interested. In the Russian-occupied zone, Red Army technical intelligence teams discovered the Sänger-Bredt report in the library of a German scientist who had not been able to flee in time.

The report was read by a Soviet expert who wrote a digest in Russian which was passed up the chain of command and eventually landed on the desk of Joseph Stalin. Outwardly affable when dealing with leaders of the Western Allies, the five-foot-six Stalin was cunning, ruthless, and mentally tough. At the Yalta conference at which the Big Three agreed to slice up vanquished Germany into four parts, Stalin casually remarked to President Franklin Roosevelt that he intended to "execute 50,000 officers of the German Army."[3]

Stalin was intrigued by the Sänger-Bredt report and ordered a full translation to be made. Shortly afterward, the Soviet leader called a conference to discuss the report. Among those present were Vassili Stalin, son of the *voshd* (boss); Vyacheslav Molotov, commissar of foreign affairs and a bitter foe of the United States; Georgi Malenkov, Stalin's private secretary and a power in the Communist party; and Lavrenti Beria, chief of the Soviet secret police.

Several technical experts were present, one of whom had closely inspected the underground factory in the Harz Mountains, so he began discussing the potential of V–2s. His recital was cut short in mid-sentence when Georgi Malenkov snapped: "The range is too short. Who do you think we are going to fight—Poland?"[4]

Stalin seemed to agree that time was being wasted talking about a missile with a range of perhaps 160 miles. Such a missile would not even reach Paris or London from the Soviet Union, much less New York City and Washington, D.C. So the premier ordered another scientist to deliver a lecture on the potential of the antipodal, rocket-assisted bomber in the Sänger-Bredt report. Intrigued, Stalin wanted to know the whereabouts of Eugen Sänger. Secret Police Chief Beria had to confess that he did not know. "Then have somebody find Sänger and bring him to the Soviet Union in a voluntary-compulsory manner," the dictator ordered.

Vassili Stalin, a pilot in the Soviet air force, would undertake the hunt. Flying to Berlin in strict incognito, he called at the headquarters of Colonel General I. A. Serov, the Russian commandant in Berlin. Serov sent for Serge Tokaev, his senior scientific advisor, and directed him to join with Vassili Stalin in tracking down Sänger. If Sänger refused to sign a contract, he was to be kidnapped, Serov declared. "Remember," he added, "Comrade Stalin relies on you to produce results!"[5]

Lieutenant Colonel Tokaev and young Stalin spent many weeks searching for Sänger, first in Berlin, then in Paris. Reports floating back to General Serov in Berlin were that the two Soviet investigators were spending more time in German and French nightclubs and assorted shady dives with shady women than they were devoting to locating their quarry, Sänger.

Finally, Tokaev and Vassili Stalin reported back to Serov that they had failed in their mission. The elusive Sänger was surely dead or deep undercover somewhere in Western Europe.

Their failure was a curious one. Sänger and Irene Bredt, his assistant and mathematical whiz, had since married and were living openly, with no secrecy, under their real names.

When U.S. investigators began looking for the professor, they had no trouble in locating him. Sänger was interrogated at length, proved to be cooperative, and volunteered to give two copies of his antipodal bomber report to the Americans. Later, the lengthy document was translated by the U.S. Navy under the title "A Rocket Drive for Long-Range Bombers."

On June 28, 1945, Colonel Holger Toftoy was in Brussels when he received an urgent order to report at the earliest possible time to Lieutenant General Levin Campbell, the chief of ordnance, in the Pentagon. Toftoy was puzzled. Why would a three-star general want a colonel to fly from Paris to Washington, especially on such short notice?

When Toftoy walked into General Campbell's office, the colonel was in for a pleasant surprise. In line with War Department procedures, Colonel Gervais Trichel was being transferred overseas, and Toftoy would succeed him as chief of the Ordnance Rocket Branch. Toftoy was elated. By serving in the Pentagon, he would have the opportunity to accelerate the approval of a War Department plan to import key German missile experts.

Toftoy quickly learned that a behind-the-scenes debate was being waged in official Washington over the wisdom of the Pentagon plan. Pro and con voices were heard in the War, Justice, Commerce, and State Departments. Most of the American scientific community was stridently opposed, including Howard P. Robertson in the Field Information Agency, Technical (FIAT) in Europe. "In allowing the Peenemünde boys to continue their developments, we are perpetuating the activities of a group which, if ever allowed to return to Germany or to communicate with Germany, can in fact contribute to Germany's ability to make war," Robertson declared in a memorandum. Another American scientist expressed the fear that the German missile experts would organize "underground cells" for the purpose of spreading propaganda to "foment an uprising" against the U.S. government.

In the White House, President Harry S Truman was being bombarded by appeals from both sides. A friendly, blunt Democrat from Independence, Missouri, Truman had become president on April 12, 1945, when Franklin Roosevelt died suddenly 83 days into his unprecedented fourth term.

Late in June 1945, President Truman was seated behind his polished mahogany desk in the Oval Office poring over a lengthy memorandum from John Franklin Carter, Jr., who had been an unofficial intelligence advisor to Franklin Roosevelt during World War II and now was filling a similar role for Truman. Under the pen name Jay Franklin, he was a widely known syndicated newspaper columnist, author of several books, and a radio commentator on the National Broadcasting Company (NBC).

Prior to his embarking on a career in mass communications, Carter, a graduate of Yale University, had worked in U.S. embassies in Rome and Constantinople, as well as in the office of the secretary of agriculture in Washington. His memo said:

> The attached two [technical] reports constitute a confession of absolute scientific defeatism on the part of both the [U.S.] Office of Scientific Research and Development [headed by Vannevar Bush] and the War Production Board's Office of Research and Development.
>
> American armies have captured a number of top German scientists who constitute part of our "loot." Many of the Germans are willing to come here and work under any terms laid down for them. However, our top scientific officials say that none of them should be brought here for fear of arousing professional jealousy. Instead, they propose to lock them up in France and say, "Work!"
>
> The fact is that the Germans are years ahead of us in missiles, aerodynamics and many other technical fields. . . . Our scientists' faces might be red and they might be discredited if we brought other Einsteins, Sikorskis and Steinmetz's to this country.[6]

Five days later, John Carter wrote Truman again, this time taking issue with the findings of a committee formed by Vannevar Bush. His panel was charged with the task of "investigating the entire scientific and research sector in Germany and making recommendations on postwar United States policy."

Carter told the president that Bush's committee, after spending only two weeks in Germany, "found that it would be most impractical to utilize German scientists and researchers in the United States because we have no proper places for them to work." Carter declared that "Bush's committee's finding and conclusions reflect a schoolboyish attitude toward the whole question of America's scientific and research future."[7]

On July 6, 1945, while the debate in Washington was still in progress, the Joint Chiefs of Staff, headed by five-star General George C. Marshall, approved Operation Overcast, a secret plan to bring a limited number of German scientists to the United States "for temporary military exploitation, particularly those who will assist in shortening the Japanese war."

Overcast provided for importing no more than 350 men (including 100 missile experts for Colonel Toftoy's ordnance branch), without their families. They would be paid modest salaries out of funds controlled by 75-year-old Secretary of War Henry Stimson, who Washington insiders considered to be one of the capital's sharpest minds.[8]

The plan stipulated that the German scientists would not be treated as prisoners of war but rather as free men who would voluntarily come to the United States to pursue their research. Their employment would be for a duration of six months, and the wages were set at $6 per day, plus medical care, food, and housing.

Once the final touches were put on Overcast, it had to get the approval of

Edward R. Stettinius, Jr., the 44-year-old secretary of state and former chairman of the board of the United States Steel Corporation. Pointing out that the successful prosecution of the war against Japan was uppermost in his decision, Stettinius gave Overcast the green light on July 19.[9] The landmark decision by the U.S. government was kept under wraps: it would be more than a year before the general public, and the world, knew about Overcast.

A few days after Overcast had been approved, Colonel Holger Toftoy flew to Germany to join with Richard Porter and others in selecting the 100 missile scientists to whom contracts would be offered. Negotiations were held in a two-story, European-type country house outside Witzenhausen where Wernher von Braun and other key scientists were lodged. A few years earlier, the mansion had been converted to a schoolhouse, and the rooms provided dormitories for the Peenemünde experts and their families. Beds were lined up from wall to wall. One long, continuous bed provided space for five to eight people. A small number of personal items were stacked in corridors, and one room had been cleared for Toftoy, Porter, and others to confer with the scientists.

Most of the rocket men were eager to go to the United States, for their future in chaotic Germany was bleak indeed. However, they balked over the prospect of leaving their families behind to fend for themselves in the turmoil of a battered and defeated nation. Making recruitment more difficult for the American teams was the fact that the British and French teams were promising that the scientists and engineers could bring their families with them. So the Germans were refusing to sign contracts with the United States.

At the urging of ordnance officers in Europe, General Dwight Eisenhower appealed to the Pentagon to reexamine the policy prohibiting families from coming to the United States. "Only about 1,000 people are involved," the supreme commander cabled, "which is about the only material dividend we will get from this war. It seems to me that the success of [Overcast] depends on having these scientists in a proper frame of mind."

Confronted by balking rocketeers, Holger Toftoy knew that provisions would have to be made for tending to the needs of the missile experts' families, or the major American recruitment effort would fall flat on its face. So the colonel reconnoitered the countryside and located a former German cavalry barracks northeast of Munich, near Landshut, which he thought would be adequate for housing the dependents in the absence of the men.

Toftoy hurried to Third Army headquarters and was told that the United States was not in the business of arranging comfortable quarters for German families—even those of coveted missile scientists and engineers. Refusing to accept defeat, Toftoy demanded to see the Third Army commander, four-star General George S. Patton, the fire-eating old warrior.

Patton, besieged by monumental problems in trying to bring a measure of order to this chaotic region of Germany, gave Toftoy short shrift, and a wrangle of sorts ensued between the tenacious colonel and the famed commander four ranks his senior. Patton unlimbered his ample vocabulary of profanity, but eventually Toftoy carried the ramparts after the general became convinced that the highest levels of government in Washington had approved Overcast.

American army doctors and technicians would provide medical care to the families in what was unofficially dubbed Camp Overcast. American PXs (post exchanges) and commissaries were to be available to the dependents for buying food and other necessities at discount prices. Armed American soldiers would protect the former cavalry compound.

The Third Army had the responsibility of making Camp Overcast livable, but when the first families arrived, conditions were so poor (as the Third Army would later admit) that the dependents spent their first two nights crammed into dusty railroad passenger cars parked on a sidetrack. There was no food, no American soldiers for protection, and the barracks lacked furniture, running water, and functional toilets. This situation, the Third Army cabled the Pentagon, "had undermined the confidence of the German scientists in the good faith of the American government."

Within days, the shambles at Camp Overcast were largely remedied by the Army's "selective requisition" (that is, confiscation) of large amounts of furniture from German homes in the region and by employing local plumbers to put the toilets and sinks into working order.

Meanwhile in Great Britain, Alwyn Crow, the eminent scientist who had argued vehemently that the Germans had no long-range missiles right up to the moment London was deluged with V–2s, was impatient with the lack of results British interrogators in Germany were having in wooing rocketeers into signing contracts. Crow demanded to SHAEF that key Peenemünde Germans be brought to England for interrogations before they signed up with the Americans.

Begrudgingly, Colonel Toftoy agreed, and a passel of the rocket experts, including Wernher von Braun and General Walter Dornberger, were flown to Inkpot, a school near Wimbledon that had been requisitioned as an internment camp for German scientists.

Interrogated by Crow, von Braun was cagey in his replies, for he had already made up his mind to cast his lot with the United States providing that an acceptable contract were offered. It soon became obvious that his choice was the correct one. Clearly, the postwar British missile program lacked sufficient funds, was in a state of disarray, and did not have strong support from His Majesty's government. After six years of bloody conflict, Great Britain was in no mood for the extremely costly process of developing long-range missiles. In late July, von Braun flew back to Germany.

General Dornberger remained in England. A month earlier, at Garmisch, he had been interrogated by the British through an interpreter, for the "Father of Peenemünde" did not speak English. "He should be interned as a menace to the security of the Allied forces," the Brit interrogator had written. "He is a technical member of the German General Staff with extreme views and wishes for a Third World War." As a result of such harsh condemnation, and the fact that British politicians were seeking a scapegoat for their failure to detect and halt the V–2 deluge on London, Dornberger was thrown into the London Cage, a detention center for suspected war criminals.

Hearing of the action, Colonel Toftoy was furious, claiming that this was a classic example of British double-dealing. Dornberger was badly needed for the U.S. long-range missile program, Toftoy declared. The colonel had ignored the fact that the Americans also had been guilty of double-dealing by shipping all 100 of the V–2s to the United States when, by prior agreement, the Brits should have had half of the missiles.

Since the British had been unable to locate and "give a fair trial and hanging" to the man actually responsible for directing the V–2 blitz on London, General Hans Kammler, they were determined to put some German in the docks at Nuremberg in his place. Dornberger was selected for that "honor."

Dornberger protested that he had no command responsibility for the actual firing of the V–2s and that his function had been only to develop the long-range rockets. If he were going to be executed, he said, the same should be done to all engineers, scientists, and military men in the United States and Great Britain who worked to develop new weapons. Allied bombing fleets, he pointed out repeatedly, had killed far more German civilians than had the V–2s.

Howard Robertson, chief advisor to Colonel Ralph Osborne, a scientist and chief of the U.S. Field Intelligence Agency, Technical, joined in the chorus howling for Dornberger's scalp. Robertson sent a memo to Osborne stating that Dornberger "is a dangerous man and should be shorn of all influence over and even prevented from having contact with his former Peenemünde subordinates."[10]

Halfway around the globe from Europe on August 8, 1945, a B–29 Superfortress piloted by Colonel Paul Tibbets dropped what was later described to an awed world as an atom bomb on the industrial city of Hiroshima, Japan. Casualties were enormous, but they were just a fraction of the number of Allied fighting men who would have been killed or wounded had it been necessary to invade the Japanese home islands. A week later, diminutive, soft-spoken Emperor Hirohito, the ruler regarded as a god by the Japanese, took to Radio Tokyo to announce that the empire was surrendering.

In Washington, those directing and promoting Operation Overcast were confronted by a new ballgame: the reason for bringing in German scientists (to help shorten the Pacific war) had vanished in a mushroom cloud of smoke—literally. America in a volatile postwar world was far from secure, however, and the haunting specter of a powerful Soviet Union, armed with long-range missiles developed by German experts, resulted in a high-level decision to continue Overcast.

# 14

# Threats and Blackmail

Postwar Germany had turned into a hotbed of intrigue and machinations. Although the guns had fallen silent five months earlier, an undercover, no-holds-barred battle was being fought for Germany's scientific brainpower. Secret agents for the wartime Allies—the United States, England, France, and the Soviet Union—were relentlessly poaching on one another's occupation zones in search of scientists, engineers, and technicians.

SHAEF had been dissolved, and now a steady blizzard of angry memorandums flew back and forth between headquarters of the four occupying powers. There were charges and countercharges of kidnappings, threats, blackmail, harassment, and intimidation of German experts and their families.

By September 1945, suspicion and mistrust reached the boiling point between the United States and Russia after General Lucius D. Clay, deputy to General Dwight Eisenhower, received a virtual ultimatum from General Vassily Sokolovsky, commander of the Russian zone. Sokolovsky, in strict language, demanded the return of 49 German scientists, who, he claimed, had been kidnapped by American agents from the Soviet zone.

General Clay, a native of Marietta, Georgia, had been chosen for his current post because of his reputation as a skilled diplomat, a trait that would be highly useful in the tug-of-war between the four Allied nations. At the same time, when circumstances dictated, the 47-year-old Clay could be tough-minded and resolute.

Now, in the face of the Soviet threat, Clay had to combine diplomacy and steadfastness. In a carefully worded reply to General Sokolovsky, he said that, although the 49 German experts had joined up with the Americans on their own, they would be released on October 10. In the days ahead, negotiations at the Allied Control Council (the four-power coordinating apparatus) in Berlin became so heated that Clay found reasons to delay handing over the 49 Germans.[1]

Meanwhile, the "case of the kidnapped scientists" developed even more menacing overtones when Soviet dictator Joseph Stalin wrote President Harry

Truman and demanded that the disputed Germans be handed over to the Red Army.

Harry Truman, although his facial appearance was that of a mild-mannered haberdasher, was tough as nails when the situation demanded. Truman had given the green light for dropping the A-bomb, then slept soundly that night, fully convinced that he had made the right choice and saved the lives of countless Americans. Now he had become angered by Stalin's blatant demand and privately called the Soviet leader "a blackmailing son of a bitch."

Eventually, the brouhaha over the 49 German scientists evaporated in the fog of more far-reaching hassles between the Americans and the Soviets, but the episode had driven home to President Truman how hotly contested was the struggle being waged for German brainpower. Consequently, he issued Executive Order 9604 that established a Publication Board with broad powers to obtain and make available to the American public at large and the scientific community in particular technical information extracted from the Germans. These powers also included the authority to import German scientists, engineers, and technicians to the United States in order to furnish scientific information to the armed forces.

In Great Britain, Admiral Charles Kennedy-Purvis proposed a plan to import "100 German scientists of outstanding ability" on six-months' probation, then "encourag[e] them and their families to prolong their stay indefinitely." Kennedy-Purvis' proposal promptly drew a chorus of protests. Wartime memories were too fresh, and the conviction that all Germans were dangerous was still deep-rooted among civilians and officials in the British government. The Joint Intelligence Committee (JIC) was fearful that the scientists would return home after six months with British military and scientific secrets and use this knowledge to rearm Germany.

Not to worry, Kennedy-Purvis countered, for the Germans would have to obtain specific permission to leave England, and he doubted if the British government would ever feel a need to grant such a request. Admiral Kennedy-Purvis concluded that what he needed to overcome the heavy resistance was an "alarming crisis." Forty-eight hours before he was to present his plan formally at a decisive meeting of the British Chiefs of Staff, Kennedy-Purvis got word of a bitter dispute that had erupted at Volkenrode in the British zone over whether the Americans or the Brits "owned" a prominent German aeronautic scientist, Bernard Goethert.

Lieutenant W. A. Rosenbauer, an American intelligence officer, had invaded British "territory" with the intention of grabbing Goethert, claiming that the noted designer already had been signed up to be shipped to the United States. When Rosenbauer could not locate his target, he reported to his superiors that Goethert had been "kidnapped" by British Major Roy Goody, the commander of the region. Goody had left a note at Goethert's home: "The Ministry of Aircraft Production have an interest in Dr. Goethert."[2]

Admiral Kennedy-Purvis circulated to the British Chiefs of Staff his own revised version of the Volkenrode affair: Goethert had not been "owned" by the Americans but instead had been kidnapped by Lieutenant Rosenbauer. What was more, the admiral declared, Rosenbauer had returned to the British zone and tried to abduct three more prominent German scientists who "belonged" to England. Kennedy-Purvis explained piously that the American intelligence officer's devious scheme had been foiled "only at the last moment," and demanded that Lieutenant Rosenbauer's activities in the British zone "be brought to an end."[3]

A week later, the admiral's sleight-of-hand paid off. The JIC and the Home Office had become convinced that if the British did not grab German scientists, the United States, Soviets, and French most certainly would. So a compromise agreement was hammered out that satisfied all parties and paved the way for importing German scientists and their families. As a hedge against security leaks, the Home Office was assigned the task of spying on the Germans' activities and contacts while in England.

Meanwhile, agents of the *Direction Générale des Études et Recherches de Défense Nationale* (the French equivalent to the Office of Strategic Services [OSS] of the United States and MI–6 of England) were aggressively and with much success recruiting large numbers of German experts.

A stream of reports reaching U.S. headquarters in Berlin indicated that French secret agents were playing hardball, often warning German scientists not to cooperate with American or British technical intelligence teams and recruiters. Rumors also had it that the French were arresting the wives of German scientists, engineers, and technicians who already had hooked up with the Americans or British and forcing them to blackmail their husbands into coming back to the French zone.

Meanwhile, American intelligence learned that the Russians had organized three groups to research and develop the V–2. Based in Bleicherode, in Peenemünde, and in the underground factory Mittelwerk, the collective groups were called *Rabe*, an abbreviation of *Raketenbetrieb* (Rocket Enterprise).

Probably using the Black List that the U.S. government naively had turned over to Stalin earlier, Soviet intelligence quickly rounded up the German missile experts who had chosen to remain in the Russian zone and found the man capable of directing Rabe—32-year-old, blond Helmut Grötrupp. In February 1944, Grötrupp had been arrested by the Gestapo, along with Wernher von Braun, and charged with sabotaging the German war effort, so he was deemed to be "politically reliable" by the Soviets.

Grötrupp and his attractive wife, Irmgard, had been among the trainload of scientists and their families who had been smuggled out of Nordhausen by Richard Porter and Major Robert Staver. But a short time after reaching Witzenhausen in the American zone, Helmut and Irmgard Grötrupp slipped out of camp in the middle of the night and returned to their home in Soviet-occupied Germany.

Although Grötrupp had not been one of the top experts at Peenemünde, he had held an important post as a department head, had access to secret V–2 files, and participated in countless conferences at which Wernher von Braun, Ernst Steinhoff, and other premier rocketeers spoke about missile problems and achievements.

Helmut Grötrupp became the Wernher von Braun of the Soviet missile program. In keeping with that lofty status, he and Irmgard lived a privileged life. The Russians provided them with a sizable salary, a large house, a BMW automobile with a chauffeur, servants, and hard-to-find liquors and food items. Most important to the Grötrupps, unlike colleagues who had cast their lots with the Americans, the director of Rabe would not have to leave the Fatherland—or so the couple had been promised.

In late August, U.S. intelligence picked up information indicating not only that the Soviets were going all-out to develop far more sophisticated missiles but that they hoped to snare the services of top German scientists now in the custody of the Americans. A German informant in the Soviet zone smuggled out a chilling message:

> A collaborator around Bleicherode told me Russians intend develop big rocket with range of 3,000 miles and they need specialists. Russians are pay big prices to getting over to their area Prof. v. Braun and Dr. [Ernst] Steinhoff.[4]

Proof that the Soviets were trying to entice von Braun and other key experts into their lair lent urgency to Colonel Holger Toftoy's mission of signing up the German missile aces. His recruitment goal was to bring to the United States a fully integrated missile team—scientists, electrical and mechanical engineers, chemists, aerodynamics experts, fuel specialists, and physicists. However, there was such a deep pool of scientific talent that Toftoy could not limit his list to the authorized 100 men; he instead picked 127.[5]

Among those who signed six-month contracts with a provision for an additional six months at the option of the Ordnance Corps was Wernher von Braun. As the world's foremost missile expert, he was guaranteed a salary of $750 per month, along with free medical care. The other Peenemünde men would receive commensurate salaries. Those selected were expected to put in 48-hour work-weeks in the United States, and they would pay for their food, lodging, and incidental expenses with the same $6 per diem allowance the government granted American military and civil service employees.

By late September 1945, the four Allied powers abandoned pretenses and openly engaged in a mad scramble to corral German technical know-how. The French opened a research center at Ravensburg, and the British reopened four research centers. All were staffed by German technicians. General Eisenhower authorized the reopening of five technical facilities where Germans had been employed, mainly engaged in weapons development.

Bowing to the realities, Eisenhower approved the hiring of former members of the Nazi party "as long as they had not held leading administrative and policy-making positions."[6]

At the same time, the Russians issued an order for all scientists, engineers, and technicians in their zones to register with the Soviet military government. This directive had a chilling effect on the targeted Germans: to them, it signaled the end of "voluntary recruitment" by the Russians.

On September 20, 1945, a U.S. Army transport plane touched down at New York City. Wernher von Braun emerged with six of his associates—Eberhard Rees, Walter Schwidetzky, Erich Neubert, William Jungert, Theodore Poppel, and Willi Schulze. Their entry into the United States had been arranged without formal immigration procedures by a presidential executive order, and the missile men were regarded as "wards of the Army."[7]

Meeting the vanguard of the Peenemünde team was Major James Hamill, the American officer who had smuggled the components for 100 V–2s out of the Harz Mountains ahead of the Soviets during the previous May. Hamill was now wearing civilian garb as part of the security measures cloaking the arrival of von Braun and the other rocketeers.

A few weeks earlier, Hamill had volunteered to go to the Pacific to join an infantry division. Along with a group of other officers, Hamill had been airborne for less than an hour when the pilot received a radio flash that the Japanese had surrendered. In keeping with his instructions, the pilot returned to the takeoff point on the U.S. West Coast.

A month later, Jim Hamill rejoined his old boss, Colonel Holger Toftoy, in the Pentagon and was told that he would be sent to New York to meet von Braun and the others when they landed. No doubt Toftoy recalled the strict admonition in the memorandum by Undersecretary of War Robert Patterson, who said: "I assume that these men will be under strict surveillance at all times while in the United States." Toftoy told Hamill: "Jim, I want you to stick with von Braun like glue—24 hours per day."

Hamill's "honeymoon" with von Braun was not just to prevent the escape of an enemy alien but also to protect him should someone who hated Germany learn the identity and nationality of the scientist and try to commit mayhem on him.

Von Braun's first hours in the United States were painful ones. His broken arm and shoulder had not yet healed and he was still wearing the bulky cast. Earlier in the day, he had been stricken by a hepatitis attack, which added to his misery.

Major Hamill herded his seven "wards" onto a train, and the group reached Washington's bustling Union Station shortly after dawn. Hamill and von Braun climbed down from the train, while the six other Germans, escorted by American officers, continued toward the U.S. Army's Aberdeen Proving Ground in Maryland.

Hamill and von Braun took a Yellow Cab to a leading Washington hotel, where they registered under fictitious names and settled into a two-bedroom suite. Soon they were joined by a U.S. technical intelligence team which spent the next five days questioning the missile scientist. Secretaries, sworn to secrecy and working in shifts, recorded every word.

Meanwhile, the other six Germans arrived at Aberdeen, where most of the nation's weapons and top-secret military equipment were developed and tested. They plunged into the awesome task of identifying, segregating, evaluating, translating, and cataloging the 14 tons of missile documents that had been hauled away from the Dörnten mine.

Their work was cut out for them: the mass of materials had been hastily packed in Germany and was in a shambles. However, they often could classify at a glance whether a document was important or trivial and how a particular piece fit into the overall scheme. They, themselves, had written many of the documents or had helped to compile them. Had the Germans not been available for this task, it would have required years for Americans to get the job done.

Back in Washington, the exhaustive interrogation of von Braun was concluded and he and Jim Hamill, now wearing his major's uniform, caught a train for El Paso, Texas, adjacent to Fort Bliss. Changing trains at St. Louis' Union Station, Hamill learned that he and von Braun were to have berths on Car O, which was filled with wounded veterans of the U.S. 82nd and 101st Airborne divisions.

Fearing that an awkward incident might erupt should the wounded GIs learn von Braun's true identity, Hamill flashed a State Department card at the ticket window and got his and the scientist's berths shifted to another car. However, the sleeping accommodations were at opposite ends of the Pullman coach, so Hamill decided that the two men would play it as though they were not traveling together but would join one another in the dining car for meals.

The Texas-bound train pulled out of St. Louis' cavernous Union Station that night, and the following morning Major Hamill settled in his seat at one end of the coach to conduct his long-range guard duty of von Braun, who was carrying on an animated conversation with a passenger at the opposite end of the crowded car.

"Where you from?" von Braun's seatmate asked.

"Switzerland," the rocket expert replied, knowing that the mountainous nation had a large German-speaking population.

The American's eyes lit up. "Switzerland!" he exclaimed. "Why, I've been there many times! What business are you in?"

Usually voluble, von Braun felt that he should say as little as possible. "Steel," he answered.

"What kind of steel?"

"Ball bearings."

Now von Braun felt that he had painted himself into a corner. He knew very little about Switzerland and even less about the ball-bearing business. Worse, his new American acquaintance knew a great deal about not only the steel business in general but the ball-bearing industry specifically.

All the while, Major Hamill, at the far end of the car, was growing increasingly nervous, fearful that the loquacious von Braun would let slip some clue to the fact that the world's foremost long-range missile expert was in the United States and bound for the White Sands Proving Ground. Worse, Hamill was racked with dark thoughts that von Braun, whose real motivation in coming to America was yet to be proved, might deliberately, for whatever reason, try to lift the cover from the intense secrecy surrounding Operation Overcast.

While von Braun was searching for means to switch the conversation, the train pulled into Texarkana, on the border of Arkansas and Texas. Much to the German's relief, his seatmate said, "Well, here's where I get off." Rising, the American shook hands with the scientist just as the anxious Jim Hamill walked up and remarked: "If it hadn't been for you Swiss, I doubt if we could have beaten those damned Germans!" Hamill struggled to keep from laughing. More important, his "ward" had passed his first loyalty test.[8]

Reaching Fort Bliss, Hamill and von Braun were greeted by the post commander, who told them that Hamill had been assigned a room in the Officers Club and von Braun one in the BOQ (Bachelor Officers Quarters). Both men welcomed this arrangement, for it would conclude their ten-day honeymoon.

That night Hamill, tired from his long spell of intense vigilance, went to bed early. He had just drifted into sleep when he was awakened by a loud banging on his door. Sleepy-eyed, he opened the door and faced the Fort Bliss provost marshal (the post chief of police, in essence) and the commandant's executive officer. As it developed, the post commander had had a change of heart and became worried about having an "enemy alien" staying at Fort Bliss without being under proper custody. "You'll have to move out of the Club and go back and live with the man you brought with you," the provost marshal declared. So Hamill tossed a few things into his valise and rejoined von Braun at the BOQ. The honeymoon was resumed.[9]

After daybreak, Hamill was confronted with a new problem: how could he pay calls to the Fort Bliss engineer, quartermaster, and other offices in order to establish his own office with a German national constantly in tow, not being able to even introduce his companion or tell why he was there? Fate settled that predicament: Wernher von Braun's liver trouble flared anew and he was admitted to the Fort Bliss hospital. Now Hamill could carry on his business of arranging for the arrival of more members of the Peenemünde team.[10]

A few days after von Braun was hospitalized, Colonel Holger Toftoy flew to Fort Bliss from Washington. After Toftoy and Jim Hamill enjoyed a meal at Lillian's Steak House a short distance from the post, they drove past what appeared to be an empty annex to the William Beaumont General Hospital, a large government facility. "That would be a great place for our setup," Hamill enthused, pointing to the annex. "It has a security fence all around it, its own fire department, even a swimming pool."

Toftoy agreed, but allowed as how he doubted if it could be pried loose from the Army's Medical Department. However, the two men decided they would reconnoiter the situation and called on the commanding officer of William Beaumont, a very senior—and very tired and harassed—medical colonel named Thomas Ryer.

Ryer declared that he was indeed having enormous headaches in keeping both the hospital and the annex running. America's huge wartime forces were being dismantled helter-skelter, and each day he was losing doctors, nurses, and corpsmen.

Toftoy and Hamill expressed the proper amount of sympathy for the Beaumont commandant's predicament and said they would try to relieve him of the burden of having to operate the annex. Ryer was delighted.

Back in the Pentagon, Toftoy called on the surgeon general and made a pitch for transferring the Beaumont annex to the Ordnance Branch. There was great urgency. Toftoy had learned that the commanding general of Fort Bliss also was casting covetous eyes on the annex for his needs, and at least two veterans' associations were seeking to get control of the building.

After a lively meeting in the surgeon general's office between involved parties, the Beaumont annex was transferred to Ordnance, effective in 30 days. On October 3, the Ordnance Corps activated the Research and Development Suboffice (Rocket), with Major James Hamill, now 27 years old, in command.

Hamill, a believer in squatter's rights, promptly moved into the annex before it became widely known what had taken place. Now there were offices, laboratories, recreational rooms, and living quarters for the main body of Germans when they arrived.

Wernher von Braun, meanwhile, had made a rapid recovery from his bout with hepatitis, and with time on his hands, while awaiting the arrival of the main body of the Peenemünde team, his thoughts once again turned to space travel. Mars, named after the ancient Roman god of war because of its fiery red color, especially intrigued him.

As the days passed, von Braun designed a passenger space capsule, based on his own mathematical calculations, that could land on Mars, 35 million miles away. Space explorers would take along a truck for driving around Mars, on which vegetation was thought to grow at certain times of the year. Von Braun's scheme was for earthlings to blast off in the rocket-propelled capsule, refuel halfway at a space depot, land on the red planet, and construct underground air-conditioned caves in which to live.

# 15

# Rattlesnakes, Tarantulas, and Guided Missiles

On the drab, cold morning of November 16, 1945, the gray-painted USS *Argentina*, an American troop transport, sailed into New York harbor. On board, along with hundreds of returning GIs, were 55 members of the Peenemünde missile team. Their arrival was cloaked in secrecy, but the next day the *New York Herald Tribune* reported that a group of "Reich citizens" had been brought to the United States to drive trucks for the Army Transportation Corps. "All the Germans seemed to be quite jovial as they walked down the gangplank to waiting Army buses," the brief article said.

Escorted by Army officers, the 55 "truck drivers" went by rail to Fort Strong, an island in Boston harbor, headquarters of the U.S. Counterintelligence Corps. There, the Germans were screened for two weeks before climbing aboard a train bound for their new temporary home at Fort Bliss, Texas, where they would rejoin Wernher von Braun.

In the meantime, Colonel Ernest Gruhn, an energetic, zealous officer, had taken over the direction of Overcast from General Clayton Bissell, the intelligence officer for the Joint Chiefs of Staff. As the new head of the Joint Intelligence Objectives Agency (JIOA), under the Joint Chiefs, Gruhn's function was to promote the recruitment of German scientific experts while denying their know-how to the Russians.

Working in a suite in Washington's old Munitions Building, Colonel Gruhn immediately grasped that the U.S. government had become paralyzed over the delicate question of importing large numbers of German scientists. Gruhn learned that Overcast was being reevaluated "with a view of reducing to a minimum the numbers of key [German] experts" to be imported.[1]

Undaunted by the apparent mood in certain high councils of the Truman administration to emasculate or scuttle Overcast, Colonel Gruhn set out to enlist the affirmative support of influential figures in Washington, one of whom was Henry A. Wallace, the secretary of commerce. Born on a farm near Adair, Iowa, in 1888, Wallace had served as vice president under Franklin Roosevelt from 1941 to early 1945, when he was replaced on the Democratic ticket by Senator Harry Truman of Missouri. Wallace had not been renominated at the Democratic convention because many powerful party leaders did not cozy up to his far-left idealism.[2]

Despite his liberal leanings, Wallace was sold by Colonel Gruhn on the benefits to American commerce that would accrue through the importation of German brainpower. On December 4, 1945, Wallace wrote a letter to President Truman:

> The transfer of outstanding German scientists to this country for the advancement of our science seems wise and logical. Presently under our control [in Germany] are eminent scientists whose contributions, if added to our own, would advance the frontiers of scientific knowledge for national benefit.
>
> Many of the German scientists have already been transported to Russia . . . where their past and future knowledge will be incorporated in the scientific endeavors of that nation. . . . It is evident that many of the outstanding German scientists will no longer be available unless a decision is made quickly to permit their importation to this country.[3]

Secretary Wallace concluded his letter by ''urgently'' recommending that President Truman declare the importation of German scientists to be official U.S. policy. The Departments of State and Labor, along with the Publication Board (created by Truman four months earlier to disseminate German technical knowledge in the United States) could ''work out practical methods for bringing these men to this country,'' Wallace added. ''Intellectual reparations may well be the most enduring national asset we can obtain from the prostrate German nation.''

Harry Truman was impressed by Wallace's logic, scrawling a note on the memo: ''Interested in suggestion. Have it looked into. Apparently no law against it.''[4]

While official Washington continued to bob and weave behind the scenes, with no one wanting to be given ''credit'' for bringing in large numbers of scientists from a country that had so recently been a bitter enemy, the 55 German missile experts that had landed in Boston arrived at Fort Bliss in mid-December 1945. It was a youthful group: their average age was 33.

The youngest of the lot was 25-year-old Walter Wiesman, who had been drafted into the Luftwaffe as a teenager in 1939, then had been transferred to the Peenemünde missile experimental center on May 1, 1943. Many years later, Wiesman recalled those early days at Fort Bliss:

> Our Peenemünde team was assigned a rather rundown cottage near our quarters in the former Beaumont Hospital annex and was told it was intended for our recreational needs. All of us, including Wernher von Braun, pitched in to repair and spruce up the cottage, which we called the Clubhouse.
>
> There we played chess and cards and talked shop during our off-duty hours. The Club's ''grand opening'' was on Christmas Eve 1945. I was the bartender, and business was brisk.[5]

Since Overcast had been launched in Germany in the spring of 1945, it was the policy of the American officers involved to treat the German scientists with courtesy and tact on the theory that no one could be forced to reveal his knowledge and work to apply it for the benefit of the United States. So on the arrival of the Peenemünde team at Fort Bliss, Army officers handed each member a diplomatically phrased printed sheet that explained why the Germans were there and the conditions under which they would live and work:

> We would like you to know and to appreciate that you are here in the interest of science and we hope that you will work with us in close harmony to further develop and expand upon subjects of interest to ourselves as well as to you.
>
> A [U.S. Army] officer has been assigned to your quarters who will see to it that you are not molested by unauthorized personnel, and that your food is of good quality. You are not POW's but are more in the category of employees of the United States.
>
> It is also perhaps well to point out to you that you will see a guard posted at the gate (of the annex) and perhaps a military policeman in the immediate vicinity of your quarters. We want you to know that these people are there, not to confine you, but to protect United States Government property. . . . In short, do not think of yourself as under restrictions while here.[6]

Privately, Wernher von Braun and the other Germans joked with great glee over the official explanation for the armed guards and military policemen. ''We aren't POWs,'' the irrepressible von Braun chuckled. ''We're POPs—prisoners of peace!''

U.S. authorities also were careful to assure the Germans that their mail back to Germany was not going to be censored. ''However, it will be appreciated if you will kindly leave your personal letters unsealed on my desk for routine casual inspection, which will be a mere formality and is in no way intended to check your mail,'' an American officer assured the Germans.[7]

Von Braun and his associates were cautioned that the V–2 program had been classified as top-secret by the War Department and that they were not to discuss their work or reasons for being at Bliss with anyone other than duly authorized personnel. Any violation of these security regulations would be treated as though a U.S. government employee or military person had committed the indiscretion.

Special efforts were made to protect the Peenemünde superstar, Wernher von Braun. His English was far from fluent, and it was feared that he might be contacted by a foreign spy and inadvertently drop secrets. So he was given a card to carry in his wallet. It said that if he became separated from his military escort, ill, injured, or lost, he was to be asked no questions and the military officer at his quarters was to be notified immediately.

Other security problems provided Army officers with king-sized headaches. Civilian employees or Army enlisted men could not be used to operate the Germans' mess hall, because the missile experts constantly talked shop while dining. So this potential leak was plugged by assigning a contingent of German POWs to cook and to wait tables.

It was impossible to ''hide'' nearly 120 people, so the presence of the mystery men in the annex was widely known within the medical staff at adjacent William Beaumont Hospital. Speculation was rampant. Who were these men? Why were they here? Why the armed guards? Army security officers sought to curb the conjecture by floating rumors that the men in the annex were engaged in secret research on advanced medical equipment.

Other potential sources of security leaks were the American military personnel who, by necessity, were assigned to duty in and around the annex. So a set of rules and regulations was set up for the GIs. They were not permitted to drink alcoholic beverages on the premises, conversation with the Germans was prohibited, no one was allowed to enter the living quarters of the scientists except to fulfill assigned duties, relatives and friends were not allowed on the premises, and anyone found intoxicated would be dealt with severely.[8]

Like a well-oiled machine, the Peenemünde team rapidly plunged into its task—inaugurating the U.S. V–2 program, a plan for the systematic launching of missiles at the rate of about two per month. It was hoped that the U.S. Army would gain experience in the handling and firing of huge missiles, obtain information for the design of longer-range and more sophisticated missiles, gain ballistics data for building ground equipment to track trajectories and to measure missile speeds, and to research the upper atmosphere.

Their ''workshop'' was the sprawling, bleak White Sands Proving Ground, covering 2,226,013 acres, in south-central New Mexico, 60 miles north of Fort Bliss. Almost as large as Connecticut, the missile range extends for 120 miles from north to south and 40 miles from east to west.

White Sands is the largest pure gypsum desert in the world; into this dry lakebed Mother Nature had packed enough gypsum to make wallboard, plaster of Paris, and porcelain to supply the world for thousands of years. But not a single grain could be used commercially, for the federal government had declared White Sands a national monument.[9]

Summer temperatures reached in excess of 120 degrees Fahrenheit by day and sometimes plunged to near freezing at night. Heavy gusts of wind often filled the rocketeers eyes, ears, and throats with fine, white powder, making breathing and seeing difficult. Only such creatures as coyotes, field mice, lizards, and rabbits could take a steady diet of this harsh environment and survive. A few clapboard shacks had been thrown up on the Proving Ground for field laboratories and administrative functions, and the Germans often had to evict tarantulas and rattlesnakes before work could be conducted.

Problems confronting von Braun and his associates were vast and perplexing. When they opened the crates holding the V–2 components that had been shipped from the Harz Mountains of Germany, they discovered, to their dismay, that the packing had been done in such haste that not only were the parts jumbled together but many were bent and rusted.

At Peenemünde, the rocketeers always had had the finest tools and equipment at their disposal. Now they had to make do and improvise, often using equipment that had not been designed for the task at hand. When they had to convert an old trailer into a test stand for holding missiles, Wernher von Braun, always the optimist, insisted that it was a challenge to their ingenuity.

The missile men needed low atmospheric pressure to conduct certain experiments, but the necessary huge wind tunnels or buildings, such as they had in Germany, did not exist. So they towed their experimental equipment up a New Mexico mountain to an elevation of 10,000 feet for the necessary conditions.

Management of the V–2 program was divided. The Proving Ground was under the command of an Army officer, Lieutenant Colonel Harold R. Turner, and Ordnance was responsible for the test-firings. General Electric, as part of its Project Hermes, was in charge of readying the missiles. Various scientific institutions, government agencies, and private universities would furnish the instruments to be carried aloft by the V–2s, and the overall operation was coordinated by the Naval Research Laboratory. Willi Ley, a famed German rocket pioneer who later came to the United States, recalled: "It all looked a bit complicated to an outsider, but it probably was as simple as could be managed."[10]

In late January 1946, Colonel Holger Toftoy flew in from Washington to see for himself how the V–2 program was progressing. In a little black book, he kept track of the Germans' problems and needs. One request he heard time and again was that of relaxing some of the rules. Although they were in military custody, Toftoy replied, "It's not as though you are prisoners of war, being marched around by a soldier with a loaded gun." "Herr Colonel," Wernher von Braun remarked, "the only difference is that the gun isn't loaded."[11]

Meanwhile in Washington, Secretary of Commerce Wallace had grown agitated over the failure of President Truman to reply to his letter, written seven weeks earlier, in which Wallace proposed making it official policy to bring in German experts, integrating them into assorted scientific activities. So on January 18, 1946, Wallace sent a note to Matthew J. Connelly, secretary to Truman:

> I wonder if you would be so good as to ascertain to whom the President referred my earlier letter or what other action may be forthcoming. This is a rather vital subject in connection with . . . exploiting German technology for our own uses.[12]

Matt Connelly checked with his boss, then telephoned Henry Wallace that the president was still considering the matter. What the aide did not tell the commerce secretary was that influential figures in the administration were urging against the admission of German scientists and that the issue had become a political hot potato.

President Truman had sent Wallace's letter to Vannevar Bush, director of the Office of Scientific Research and Development, for his analysis and recommendation. On January 22, 1946, Bush replied, stating, "I question the soundness of such a policy of stimulating the immigration of German scientists for permanent residence in the United States.''[13] Bush added, ''I feel strongly that the proper cure for this situation [lack of sufficient American scientists] is to release technically trained Americans who had been inducted into non-technical work in the armed forces, and to resume the intensive scientific training of promising young American men and women.''[14]

Two days later, Harry Truman responded to Bush with the type of barb for which the brash president from Missouri had already become noted: ''I was morally certain that our home boys would not want any competition.''[15]

Although high officials in the U.S. State Department were continuing to voice deep concern over the wisdom of bringing large numbers of German technical experts to America, intelligence reports flowing into Washington from Europe caused Truman administration officials to worry about the Soviets cornering the German brainpower market. So in late February 1946, Secretary of State James F. Byrnes, a Charleston, South Carolina, native, and Secretary of War Robert Patterson hammered out a new unofficial policy for Overcast.

Byrnes and Patterson agreed that importing German scientists for only six months would result in their returning home and conceivably selling their know-how (and U.S. military secrets) to the Soviets. They proposed long-term employment and eventual American citizenship. Keeping the German scientists out of the hands of the Russians would be possible only if their families were allowed to immigrate to the United States, the two cabinet members concluded.

A few days later, on March 4, the Byrnes-Patterson agreement was approved by the State-War-Navy Coordinating Committee (SWNCC). It declared that entry to the United States would be denied to any German technical experts or members of their families ''who were active Nazis or otherwise objectionable.''

Now that the door had been opened for the long-range exploitation of German scientific brains, the responsibility for implementing the expanded recruitment policy fell on the shoulders of General Joseph T. McNarney, a Pentagon planner, who had succeeded General Eisenhower as U.S. military governor in Germany. Washington ordered McNarney to recruit 1,000 German scientists whose talents would be in the ''national interest,'' to care for and provide extra rations for their families, and to prevent the Soviets from contacting German scientists in the American zone.[16]

McNarney was far from happy over the new burden. He doubted if there were anywhere near 1,000 scientists in his zone who would meet the Washington criteria. And how could he prevent undercover Soviet agents in civilian garb from making contact with German scientists?

Meanwhile in the United States, the presence of some 180 German scientists—particularly those at Fort Bliss and at Wright Air Force Base near Dayton, Ohio—was known to large numbers of Americans in those regions, but the remainder of Overcast continued to be classified top-secret. However, the potential for a major security leak had arisen. The scientists had been writing to their families and addressing the envelopes in care of Camp Overcast, the unofficial name for the old cavalry barracks near Munich. U.S. Army censors had been ordered not to prevent the letters from being delivered for fear that such an action would unmask the entire operation.

So on March 13, 1946, the Joint Chiefs of Staff changed the operation's code name from Overcast to Paperclip. The new designation resulted from the fact that American recruiters in Germany fastened paperclips to the file cards on which were written the names of scientists to whom contracts would be offered to go to the United States.

In the parched desert of New Mexico, the German scientists were not allowed to participate directly in the guided-missile program because they were not U.S. citizens. However, with Wernher von Braun and the other Peenemünde men ''unofficially'' providing direction, the first V–2s were assembled, and the initial static test was conducted at White Sands on March 14, 1946. Then, on June 28, another V–2, fully instrumented for upper-air research, was launched and soared over the nearby 5,000-high Organ Mountains and 67 miles into space. Unbeknownst to most observers, America's first tiny step had been taken in what would be a 22-year odyssey to the moon.

Five thousand miles from White Sands on October 22, 1946, Helmut Grötrupp, Wernher von Braun's long-time associate at Peenemünde, and his key officials were summoned to a conference in Nordhausen by General Iznov Gaidukov, the Soviet overseer of Institut Rabe. All day, a tedious discussion was held on the future of the Russian guided-missile program. At dusk, General Gaidukov invited the Germans to be his guests at a banquet that night, and although the exhausted scientists would have preferred going home, none dared to refuse. Vodka flowed copiously and the banquet turned into a lively affair that continued long after midnight. None of the Germans had an inkling that the day's entire scenario was a carefully orchestrated machination by the Soviet Secret Police.

# 16

# Operation Osvakim: A Mass Kidnapping

Helmut Grötrupp's wife, blonde, attractive, 28-year-old Irmgard, had never adjusted to being a "rocket widow." Like the wives of the other experts in the German Rocket Collective (as the Soviets called the scientists), Irmgard harped constantly about her husband's obsession with technical problems. "Solving a mathematical equation means more to Helmut and the other men than having a comfortable bed," she complained.

These rocket scientists were so immersed in their technology, Irmgard Grötrupp told a friend, that "they don't even care what's on the plate served them at dinner." "Take Wernher von Braun," she said. "He eats his potatoes in their skins so he could save himself the time it would take to peel them."

At *kaffeeklatschen*, Irmgard told other wives that her husband had no need for alcohol or narcotics. "When some new development is discovered, he walks on air," she explained.

Other wives had similar complaints. One of them declared that her husband was more exhilarated over solving a complicated mathematical equation than he was in going to bed with her.

As a "rocket widow," Irmgard never grew adjusted to her husband dashing in and out of their home at all hours of the day and night. So when Helmut Grötrupp had not returned from General Gaidukov's bash by 1:00 A.M., she went to bed in an angry mood.

Just over three hours later, Irmgard was awakened by the rumble of trucks and rushed to the window. A searing streak of fear pierced her being. Soviet soldiers, armed with rifles and submachine guns, were leaping out of the trucks, which were parked in a circle around her house.

Moments later, there were ham-fisted poundings on the front door and the doorbell buzzed incessantly. When Irmgard shakily opened the door, Red Army men bolted past her. On their heels, a grim Soviet captain simply advised her that the Grötrupps were being taken to Russia—immediately.

The Soviet soldiers ranged around the house with crates and other containers, and in less than an hour they had stripped the premises bare. Numb with cold and fear, Irmgard was bundled into a car and driven to the Nordhausen railroad station, the same spot from which the Grötrupps had boarded the train that had whisked 1,000 German rocket men and their families to the American zone in May 1945.[1]

Irmgard Grötrupp's ordeal coincided with the launching of Operation Osvakim, the most far-reaching and bizarre mass kidnapping in 20th-century Europe. All through eastern Germany that night, Colonel General Ivan Serov's Soviet Secret Police and battalions of Red Army soldiers struck with swiftness and military precision. Armed with lists of those Germans who might help the Russians surpass the United States in missiles and in technology, soldiers sealed off entire cities (including East Berlin) while arrest squads systematically combed houses and apartment blocks.

Unlike the situation facing the Americans, the Soviets were not inhibited by the necessity of getting voluntarily signed contracts, nor were the targeted German experts screened for possible high-level involvement in the Nazi party. Entire families, from infants to aged grandparents and even pets, were rousted out of bed, hustled into waiting trucks, and driven to railroad stations. In some instances, wives were left behind and mistresses brought along instead.[2]

Many of those being abducted were allowed 30 minutes in which to pack what belongings they could carry. Those who protested too violently were strong-armed out of their homes without even a change of clothing.

That night and during the day, some 20,000 Germans were caught in the Soviet dragnet, including Helmut Grötrupp and 198 of his key associates, along with hundreds of other missile experts assigned to the test station at Lehesten. When Grötrupp complained about the mass abduction, he was told that if he did not want to be involved in missile research for the Soviet Union he would be taken to Russia's Ural Mountains to work as a laborer in the mines.

Snow was pelting drab, gray Moscow on October 27, 1946, when the first of 92 trainloads of kidnapped Germans began to arrive. After remaining on sidings for several days, the trains headed for their destinations around the vast expanses of the Soviet Union. A large number of missile experts were sent to Gorodomilia, 190 miles northwest of Moscow, while Helmut Grötrupp and his key men from Nordhausen remained in the Soviet capital.

Grötrupp, as the German leader of missile research and development for the Soviets, and Irmgard were given preferential treatment. They were assigned a large *dacha* with a staff of Russian servants in the prosperous suburb of Datschen, where Soviet stage and screen stars and prima ballerinas lived in luxury. Soon the Grötrupps' BMW automobile was shipped in from Nordhausen.

Grötrupp protested working conditions to Colonel General D. F. Ustinov, minister of the defense industry, asking how the German experts could be expected to accomplish anything when they had no tools or equipment with which to work. Ustinov was unaccustomed to being confronted with sharp criticism of his ministry, even by Soviet officials, much less by Germans. He reminded Grötrupp that he and the other rocketeers had been brought to Russia to reconstruct the V–2, not to harp on conditions.

"Then when will we be allowed to return to Germany?" Grötrupp asked. General Ustinov threw back his head, laughed uproariously, and replied: "When you can orbit the world in a missile!"[3]

Despite the disorganized beginning, Grötrupp and the German Rocket Collective soon were working in vastly improved conditions. New equipment arrived, the old factory was modernized, and the reconstruction of the V–2s, using components brought from the underground factory in the Harz Mountains, made steady if not spectacular progress.

Cries of outrage spread across the globe in the wake of Operation Osvakim. The Kremlin made no direct response to the torrent of condemnation but used the tightly controlled Soviet press in an effort to blunt the criticism.

The Red Army newspaper, *Taegliche Rundschau*, declared that the world press had "vastly distorted" accounts of what it termed the "mass transfer" of Germans who had volunteered to go to the Soviet Union.[4] The *Berliner Zeitung* charged that the United States had abducted at gunpoint some of Germany's leading rocket scientists and had taken them against their will to the United States. What was more, the *Zeitung* said, the United States and England, in violation of prior agreements reached by the Big Three, had secretly conspired to divide German scientists and deny them to the Russians. "More than 2,400 German experts are being held prisoners in the American and British zones of Germany," the newspaper said. "Most of them are dedicated Nazis."[5]

In Washington, the Truman administration concluded that now was the ideal time to remove the veil of secrecy from Overcast and Paperclip. On December 3, 1946, the Public Relations Division of the War Department issued a press release detailing the reasons for bringing over the German scientists. "After a year of experiment and research, wider avenues of research have been opened to them," the press release said. "They have proved themselves technically and morally, and it is now planned to increase the present number of 270 German technical experts to approximately 1,000."[6]

The accomplishments of the Germans in numerous fields was ticked off. Not only had the 118 rocket experts at Fort Bliss saved up to ten years in U.S. research by showing American scientists blind alleys to be avoided and fields that should not be explored, but the use of the Germans' knowledge and experience saved at least $75 million in missile research, the War Department emphasized.

What the War Department could not reveal was that the United States and Russia, the only nations to emerge from World War II as superpowers, were embroiled in a hectic, no-holds-barred battle for missile and technological supremacy.

Hardly had the ink dried on the War Department press release than a torrent of protests erupted in many quarters across the land. A year and a half after

Victory in Europe Day, animosity toward Germans still ran deep. Newspaper editorials demanded answers. Why should a German named Wernher von Braun be looked to for advice and guidance in rocketry? Was there not already a first-rate American rocket program? Since the United States had won the war, did that not prove that American scientists and engineers were more competent than the Germans?

Von Braun was the target of the most bitter attacks. "Look at this man," thundered one editorial. "He is the one who lost the war for Hitler. His V–2 was a great engineering achievement, but it had almost no military effect, and it drained German brains and material from more practical weapons."[7]

In the White House, letters and telegrams of protest arrived by the dozen. On the night of December 4, only a few hours after the press release was issued, a telegram came from C. Montieth Gilpin, who identified himself as the president of "The Society for the Prevention of World War III," in New York City. Gilpin declared that the importation of 1,000 German experts would result in the creation of a German fifth column (a spy underground) in the United States and jeopardize American security "to a frightening degree."[8]

Officials at America's elite universities, prominent Protestant clergymen, leaders of Jewish groups, and the heads of a few labor unions marshalled forces to condemn the importation of German scientists and their families. On December 30, 1946, the White House received a telegram from James Wadell Alexander, president of Princeton University. It was co-signed by 23 influential Americans, including Albert Einstein; the Reverend Norman Vincent Peale; Rabbi Stephen S. Wise, president of the American Jewish Congress; Philip Murray, president of the CIO (Congress of Industrial Organizations) labor organization; and A. Philip Randolph, president of the Brotherhood of Sleeping Car Porters. Alexander's telegram said:

> As American citizens, permit us to express our profound concern over reports that Nazi scientists have not only been brought to this country by the United States Army for research projects but that their families are to follow them and that they may be permitted to remain here permanently.
>
> We hold these individuals to be potentially dangerous carriers of racial and religious hatred. . . . This raises the issue of their fitness to become American citizens or hold key positions in American scientific, industrial and educational institutions.
>
> If it is deemed imperative to utilize these individuals in this country, we earnestly petition you to make sure that they will not be granted permanent residence of citizenship in the United States with the opportunity which would afford inculcating those anti-democratic doctrines which seek to undermine and destroy our national unity.[9]

W. A. Higenbotham, executive secretary of the Federation of American Scientists (which claimed 3,000 members), wrote to President Truman and "demanded" that "imported Germans" be denied jobs in private industry or education. "Any favor extended to such individuals, even for military reasons, represents an affront to the people of all countries and to our unfortunate scientific colleagues of formerly German occupied lands,'' Higenbotham declared.[10]

On Capitol Hill in Washington, some member of Congress took to the floor to denounce Paperclip bitterly. Representative John Dingell, a Democrat from Michigan, thundered, ''I never thought America was so poor mentally that we would have to import for the defense of our country.''

By happenstance, the War Department press release was issued on the same day, December 3, that the first 15 families of German scientists arrived by ship in New York. Major General S. J. Chamberlain, G–2 for the General Staff, ordered that the dependents be kept aboard overnight ''in order to avoid reporters and prevent embarrassment'' to the newcomers. Four days later, the vanguard of the German families reached Fort Bliss and were greeted with a rousing and tear-filled welcome by the Peenemünde men.

By March 1, 1947, there were some 400 German scientists, engineers, and technicians in the United States, most of them at Fort Bliss and Wright Air Force Base. In the eyes of the Pentagon, these ''wards'' had become even more crucial to national defense, for the Cold War between the West and the Soviet Union and its allies had grown more intense and threatened to erupt into a hot, or shooting, war.[11]

Against this haunting backdrop, gaining legal status in the United States for the German experts was a priority goal of the Joint Intelligence Objectives Agency, the group in charge of Paperclip. Many of the imported scientists, their future clouded by their status as ''wards of the Army,'' were threatening to return to Germany when their contracts expired.

Navy Captain Bosquet Wev, the JIOA director, was alarmed by this situation. If the Germans were to pack up and go home, not only would their know-how be denied to the United States but, in Wev's words, ''It's a lead-pipe cinch that the Russians will grab them.''

In the interest of retaining permanent access to the skills of the German experts, it was Wev's goal (and that of the Pentagon) to eventually obtain citizenship for the ''wards'' and their families. It would be a lengthy and convoluted process. As the Germans held the status of enemy aliens, they would have to leave the United States, apply at an American consulate in a foreign country for a visa, then reenter the United States as resident aliens. Only then could they apply for citizenship.

On February 26, 1947, Bosquet Wev submitted the Germans' dossiers to the State and Justice departments as a first step toward eventual citizenship. Each dossier contained a security report on the applicant from OMGUS, the American military government in Germany, detailing its investigation of each German's background and OMGUS' judgment of whether he had been an ''ardent Nazi.''[12]

Captain Wev's hopes for rapidly gaining legal immigration status for the ''wards'' soon hit a snag. Before the State Department would consider the applications, the OMGUS security report on each one had to be approved by the governing board of JIOA. The State Department's representative on that panel was Samuel Klaus, a lawyer and career bureaucrat, who had been a bitter foe of Overcast from the beginning.

Klaus had long complained about what he said was the need for more thorough background checks of the German scientists. ''Giving these security risks legal status would violate President Truman's policy against bringing 'ardent Nazis' into the United States,'' Klaus argued vehemently.

Captain Wev declared that America's national security would benefit highly from the work of the German experts and that this consideration transcended the possibility that a handful of the ''wards'' may or may not have been involved in the Nazi party to some degree.[13]

Meanwhile, President Harry Truman had become convinced that Soviet-inspired aggression around the globe had developed into a major threat to America's security, so the scrappy bantam from Missouri adopted a ''get tough'' policy with Russia. On March 12, 1947, he spoke to a joint session of Congress and unveiled what came to be known as the Truman Doctrine, guaranteeing U.S. intervention against Communist aggression—armed, political, or propaganda—anywhere in the world.

At Fort Bliss and on the parched desert of White Sands, Wernher von Braun was plunged into deep despair. From confidential sources, he had learned that his parents, Baron Magnus and Baroness Emmy, had been driven penniless from their ancestral manor and lands and the property had been confiscated by the Communist regime. Once wealthy and influential, the baron and baroness had been herded into a cattle car jammed with other dispossessed Germans and, without funds and having only the clothes they were wearing, sent rolling to an undisclosed locale in Germany.

Wernher von Braun also was worried about the fate of Maria Louise von Quistrop, his beautiful 17-year-old cousin, who was a member of a German family involved for centuries in the upper strata of Germany's public and private life. Back when von Braun was at Peenemünde, he conducted a long-range romance with Maria Louise. They had corresponded regularly and, when duty permitted, von Braun visited her at the von Quistrop ancestral estate.

Since arriving in the United States, von Braun continued to correspond with Maria Louise, and in late 1946 he asked her by mail to marry him. She accepted, and it was agreed that the wedding would be held in Germany. This presented complications: both the Army and the State Department would have to approve von Braun's travel outside the United States.

The Army approved and State, moving with uncharacteristic speed, also gave a favorable response. Ground rules were established for the trip. Von Braun would be accompanied 24 hours per day by U.S. intelligence agents, for, as the world's foremost missile expert, he might be a tempting target for a Soviet kidnapping caper in Germany.

Von Braun and his bodyguards sailed for Europe in late February 1947, and the scientist and Maria Louise, now 18 years old, were married on March 4, in the Lutheran Church at Landshut, near Munich. The site was not far from the locale where von Braun had surrendered to the GIs of the 44th Infantry Division less than two years earlier. Among those in attendance at the nuptials were Baron and Baroness von Braun, who had been located by U.S. intelligence only 48 hours earlier and were rushed to Landshut for a joyous reunion with their son.

Few newlyweds ever had a more bizarre wedding night than did Wernher and Maria Louise von Braun. Outside their bedroom door in a cozy Bavarian lodge, two armed American plainclothes agents stood guard. Across the hallway, Baron and Baroness von Braun slept.

Broke and with no home, the elder von Brauns reluctantly accepted their son's invitation to return with him to the United States and live at Fort Bliss. It was yet another humiliating blow to the baron, who had spent a lifetime in wealth and influence. His flagging spirits were nearly crushed: now he would be listed as a dependent of his son.

Although pushing 70 years of age, the baroness, who had never known poverty or work, faced her future in a strange land in an upbeat mood. One day, she told her son that she would go to work and earn money when she reached Fort Bliss.

"What are you going to do, Momma, make rockets?" Wernher quipped laughingly.

"I will open a class for German wives and teach them English," she replied firmly.

After a three-day honeymoon in the Bavarian Alps, Wernher and Maria Louise von Braun, the elder von Brauns, and the four plainclothes agents flew to Texas. Soon Wernher plunged back into his work with renewed vigor and enthusiasm, his mind now free of worry about his parents and Maria Louise. On the White Sands range, the V–2 test-firings continued at the rate of one per month, instead of the two per month that had been initially contemplated.

As had been the case in the experimental firings at Peenemünde and at Blizna in Poland, not all of the missiles behaved. V–2 Number 14 leaped off the launchpad at White Sands; four or five seconds later, it deviated from its course and headed hell-bent southward in the direction of El Paso. Tension gripped the watching scientists as the errant missile disappeared into the clear blue sky. Visions of the 14-ton object crashing into the crowded heart of downtown El Paso danced before the eyes of all.

Just before the V–2 reached the southern end of the Proving Ground, nearly 100 miles from lift-off, the engineer operating the emergency cut-off managed to crash the missile harmlessly into the parched white desert. Collective sighs of relief were issued.

Fears among the rocketeers that the 120 miles by 40 miles range was too small soon proved to be justified. At 7:00 P.M. on May 29, 1947, a V–2 rose from the launching pad routinely, but after reaching a high altitude, it suddenly shifted course and raced southward. High in the heavens the V–2 soared over the Proving Ground boundary, over El Paso, across the Rio Grande River, and on toward Juárez, Mexico.

At that moment, huge throngs were participating in a gala fiesta in the center of Juárez, unaware that a huge rocket was winging toward them. Tequilla flowed, bands played, voices were raised in song, and dancers gyrated.

Fortunately, the V–2 crashed into an uninhabited hill a mile south of Juárez. A considerable amount of residual fuel must have been in the tanks because the impact was followed by a loud explosion that rocked the region for miles but failed to halt or even slow the merriment in downtown Juárez. What could have been a messy international incident was avoided, partly due to the fact that there had been no Mexican casualties or property damage. When the commanding general of Fort Bliss rushed across the Rio Grande bridge to Juárez to extend the U.S. government's apologies, Mexican authorities were gracious and brushed off the episode as of no consequence.

A month after the errant V–2 plunged into Mexico, Wernher von Braun received exciting news. In early July 1947, General Walter Dornberger, his longtime friend and technological associate, had quietly been released from prison in England where he had been held for two years. No charges had ever been filed against him. The former Peenemünde commander returned to U.S. custody in Germany, where he told confidants that the reason he had not been tried for the V–2 attacks on London was because American engineers and scientists who had built the atom bomb dropped on Japan also would have to be tried.

Dornberger hoped to rejoin von Braun at Fort Bliss, but the War Department turned thumbs down. Public opinion in the United States might be aroused by placing a German general in control of civilians. So Dornberger signed a contract with the U.S. Air Force as a guided-missile consultant, and, without fanfare, he took up residence at Wright Field in Ohio.[14]

Operation Paperclip (originally Overcast) officially was concluded on September 30, 1947. In a press release, the U.S. Department of Defense said that 457 German scientists (including 118 at Fort Bliss) and 453 dependents had been brought to America under the program.

Hard on the heels of that announcement, the Army, Navy, and Air Force began trumpeting the benefits of Overcast and Paperclip. In New Jersey, the Army Signal Corps reported that the Germans had made "contributions of an astronomical nature" in microwave techniques, generators, and equipment design.

At its ordnance laboratory in Maryland, the Navy declared that German mathematicians and aerodynamicists had proven that "their professional educational training and theories were superior in many respects to that of United States experts."

The Air Force extolled the work of the Germans for making "astounding advances'' in jet-aircraft development, as well as in aerodynamics, airplane structures, helicopters, in-flight refueling, and a gun-sight for night fighter planes. A German-designed wind tunnel was operating at Mach 8 (eight times the speed of sound), ten years ahead of the best U.S. tunnel. ''These 209 scientists,'' the Air Force said, ''have opened up new horizons in weapons technology'' and ''they have saved the Air Force many millions of dollars and up to ten years of research and development.''

In the meantime, JIOA Director Bosquet Wev and Samuel Klaus, the State Department representative on the JIOA governing board, had been locking horns for five months over citizenship for the Germans and their families. Wev charged that the State Department was balking at approving visas because a few of the Germans were judged by OMGUS investigators as possible security risks.

Wev warned that returning to Germany the handful of scientists in question, as the State Department seemed to be demanding, in order to approve visas for the other Germans would ''present a far greater security threat to [the United States] than any former Nazi affiliations they may have had.'' Once back in Germany, these experts would be grabbed by the Soviets, who not only would benefit from their know-how but also would gain a bonanza of U.S. military secrets learned by the Germans while in America. ''By focusing on the security issue involving a handful of Germans, State is sabotaging by delay,'' Wev charged.[15]

While the Pentagon was ''pinching pennies'' (as one Paperclip scientist put it) in U.S. missile research and development, the Soviet Union was engaged in an all-out crusade to gain rocket supremacy.

Late in August 1947, Helmut Grötrupp and his associates in the German Rocket Collective suddenly vanished. Frantic efforts by fearful and tearful wives to learn of their whereabouts were stonewalled by Soviet officials. It would be two months before Irmgard Grötrupp and the other anguished wives would find out what had happened to their husbands—they had been taken into custody by Soviet Secret Police and shipped off to the bleak steppes of Kazakhstans, 130 miles east of Stalingrad.

There, in that godforsaken region, the Russians had established the Soviet equivalent to America's White Sands, and for weeks the members of the German Rocket Collective had been preparing for a crucial event—the firing of the first Soviet long-range missile, a V–2 shipped in from the Mittelwerk in Germany.

On the bitterly cold and clear morning of October 30, 1947, final preparations had been completed. With Soviet military and scientific bigwigs looking on from a bunker, the countdown began. Tension was thick enough to cut with a knife. At zero minus five minutes, the count halted: the test stand and the missile it held had slipped to one side.

Technicians ran to the stand and discovered that one leg had a broken rivet. They arduously replaced the damaged part, and the stand and missile were inched back into firing position by a huge crane. The countdown began once more. As the seconds ticked off, Helmut Grötrupp barked the command: "*Freie!*" Belching flame from its tail, the V–2 lifted slowly, surged upward, picked up speed, and disappeared into the bright wintry sky.

A raucous celebration erupted at the launchpad. Un-Russian-like, the Soviet director of the installation, beaming broadly, rushed to Grötrupp and grabbed him in a bear hug. Although no doubt pleased that the first firing of a long-range missile in the Soviet Union had been a success, the Germans were subdued. Their professional pride in the achievement was tempered by the thought that they were engaged in the development of a frightful weapon.

An hour after lift-off, the Soviet scientists and technicians erupted in another wild celebration. Word had been flashed that the V–2 had impacted directly on the target, 185 miles away.[16]

# 17

# A Fateful "Wrong Decision"

In late 1947, Pentagon generals and admirals were engaged in guerrilla warfare with one another over which service would control the development and deployment of guided missiles. Hovering above Washington during the skirmishing was a veil of deep concern over what the Russians were achieving in their rocket program. Alarming American intelligence reports disclosed that the Soviets, working with the German Rocket Collective headed by Helmut Grötrupp, had increased the range of the V–2 from 200 miles to 400 miles or longer.

Even more frightening was the report that the Russians were building an enormous missile called the T–2. It was thought to be 85 feet in length (nearly twice as long as the V–2), have a diameter of 15 feet, and be able to travel up to 1,800 miles—far enough to reach every city in Europe.

Refereeing the skirmishing between the generals and the admirals was James V. Forrestal, who, in July 1947, had become the first secretary of defense in a major overhauling by Congress of America's military establishment. A graduate of Dartmouth College and Princeton University, Forrestal had been a Navy aviator in World War I, and during World War II served as undersecretary and later secretary of the Navy.[1]

Forrestal spent countless hours listening to presentations from the Army, Navy, and Air Force and waded through a mountain of technical papers presented to reinforce arguments. Finally, the new secretary of defense, acting with the tact of a striped-pants diplomat, resolved the discord by ruling that each of the three services would conduct its own missile program, tailoring its designs to fit its own needs and goals.[2]

In private talks with confidants at Fort Bliss, Wernher von Braun anguished over whether the United States would beat the Russians into space. America was an open society, where missile triumphs and failures were plastered on front pages of newspapers and broadcast around the world. Soviet intelligence was keeping a detailed account of U.S. missile problems and goals by spending merely five cents each day for the *New York Times* and reading the complete testimony of Pentagon officials begging Congress for rocket funds.

Unlike in America where labor unions entered the picture, Joseph Stalin had no quarrels over work hours or pay: Russians labored as long and as hard and for such money as the iron-fisted dictator decreed. In the United States, members of Congress often engaged in heated debates for months before appropriating needed funds for long-range missile development. Stalin alone decided how much money he would spend to beat America in space technology.

Despite his worry over which of the Cold War combatants would win the space race, von Braun and his associates continued to conduct experiments that were helping humans take their first faltering steps into the heavens. Since the V–2s had been designed to carry a ton of cargo, they soared aloft from White Sands crammed with instruments, instead of explosives, in their noses.

Cameras were carried high into the atmosphere and photographed the earth for the first time. Other gadgets included a spectrograph to record light waves, a Geiger counter to detect radiation, gauges to register pressure and temperature, and a radio to report to ground crews. Seeds and a colony of fruit flies were sent aloft in order to analyze the effect of radiation on living cells. These were the first forms of life to soar more than 100 miles into space.

On June 11, 1948, a monkey was the lone passenger on a V–2 that gave him a speedy ride straight up for 39 miles, permitting medical specialists to assess the impact of space on animals. A rocket called a Wac Corporal was launched atop a V–2, separated from the nose of the V–2 at 20 miles up, soared to a height of 250 miles, and attained a speed of 5,150 miles per hour.

With the arrival of 1949, the Peenemünde Germans and their dependents had been in the United States for two years and the protests against their presence had started to fade. Surveillance and restrictions on their movements were relaxed. Now they drove by themselves, in their own secondhand Fords and Chevrolets, into El Paso to shop, dine, or take in the movies.

Despite their newfound freedom, the Germans remained enemy aliens. So when Iris Careen, the first child of Wernher and Maria Louise von Braun, was born on December 9, 1949, the father resolved to change the family's status to that of bona fide American citizens.

A short time later, at the prodding of the Army, the State Department came up with a plan to speed citizenship procedures. All the Germans were escorted across the international bridge spanning the Rio Grande, the border between the United States and Mexico. At the American consulate in Juárez, they filled out forms required of foreigners seeking visas to enter the United States. Their port of embarkation was listed as Juárez, and their port of entry was El Paso. Then, with visas in hand, the Germans climbed aboard old trolley cars and, for four cents, rode over the Rio Grande and into El Paso as legal immigrants.

Now the Germans were resident aliens and they applied for citizenship, a process that would require a waiting period of five years. Since Wernher von

Braun had never been in the United States officially until he rode the trolley across the Rio Grande, his four years spent at Fort Bliss did not count in his waiting period.

Von Braun, in the meantime, began speaking out on the distinct danger that the Soviets might triumph in the space race, a victory that would dramatize to the world a seeming superiority of the Communist system as compared to the American free enterprise system. He assured anyone in authority who would listen that the Peenemünde team had the capability of building missiles much larger and more powerful than the V–2 but that a vastly longer firing range was needed and more room was required for research and development.

Shaken by alarming intelligence reports that the Soviet Union had exploded its first nuclear bomb in August 1949, and that the Russians were making gigantic strides in missile development, Congress appropriated $75 million to establish a firing range at Cape Canaveral. Lying nine miles north of Cocoa Beach on Florida's east coast, the site covered 15,000 acres and was ideal for its purpose. Like Peenemünde on the Baltic Sea, Cape Canaveral was isolated and would be difficult for rubbernecking tourists—and Soviet spies—to reach by land or by sea. Instead of having a maximum distance of 120 miles (the length of White Sands), the cape would be the launching site for a 5,000-mile Atlantic Missile Range, which could extend beyond Ascension Island in the South Atlantic.[3]

Now that a suitable firing range was in the works, Colonel Holger Toftoy, Lieutenant Colonel Jim Hamill, and a few other ordnance officers began scouring the United States to fill the critical need for more research and development room. They finally set their sights on the closed-down Redstone Arsenal, which had manufactured artillery shells in World War II, and the adjacent Huntsville Arsenal, which had been used for manufacturing toxic weapons.

Located in Huntsville, a somnolent town of 16,000 people in northern Alabama, the new home of the Army guided-missile program nestled in a bend of the Tennessee River, which gave it access to the vast power resources of the Tennessee Valley Authority (TVA). Winters were mild, and Huntsville was a relatively short distance from Cape Canaveral.

Earlier, in 1946, peace had ended the Pentagon's need for Redstone and Huntsville Arsenals, and they went on the auction block. There were no buyers. So Senator John J. Sparkman, an influential figure in Washington, used his clout in an effort to have an Army Air Corps base built on the arsenals' land. Thousands of airmen would provide an enormous economic boost for the community.[4]

Senator Sparkman and Huntsville business leaders felt that the proposed air base was making headway in Washington when, like a bolt out of the blue, Secretary of the Army Kenneth Royall, on October 28, 1949, formally approved the use of the two arsenals for the Army's missile program.

Huntsville was thunderstruck: a relatively small number of German and American missile experts and their families could hardly spend as much money in the town as could droves of airmen. Even before the first contingent of Peenemünde men arrived, Huntsville, a closely knit southern city, had built up a head of steam over the imminent "invasion" of the Germans. "The last time our boys saw any Germans, they were shooting at them!" one angry townsman declared.

Huntsville, whose residents were hard-working and devout, had witnessed much of history. Founded by Revolutionary War veteran John Hunt in 1805, the town had been the site of Alabama's constitutional convention and temporary capital of the state in 1819. Huntsville felt the ravages of the Civil War when it was burned by federal troops in 1862.

Now, with the arrival of the 1950s, numerous cotton mills and a few small manufacturing plants were the main economic base, but Huntsville's principal claim to fame was reflected in its slogan, "The Water Cress Capital of the World."

In this simmering climate of hostility, on April 1, 1950, Colonel Jim Hamill, Wernher von Braun, and a few others in the vanguard of the Fort Bliss group arrived in Huntsville. In the weeks ahead, more than 500 Army personnel, a few hundred General Electric employees involved in Project Hermes, 100 government civilian workers, and 117 men of the Peenemünde team, along with their families, would trek cross-country from Fort Bliss to Alabama.

The Redstone facility was designated the Army Ordnance Guided Missile Center and would be under the command of Colonel Hamill. Although his official status remained that of a resident alien, Wernher von Braun was appointed chief of the Guided Missile Development Division—five years after he had surrendered to GIs of the U.S. 44th Infantry Division in the Bavarian Alps.[5]

During the first few months, relations were cool between Huntsville citizens and the Germans. Besides the lingering scars from World War II, and the "disaster" that resulted from the failure to obtain an Air Corps base, the Germans were viewed as being "sort of odd." Not only their accented speech but their Old World customs set the newcomers apart. Walter Wiesman, the youngest of the Peenemünde team, many years later recalled:

> Some of our older men, those in their late thirties and early forties, thought that we should avoid possible trouble by living near each other, keep to ourselves, and carry on our own activities. But Wernher von Braun and most of us felt that this was the wrong approach, that we should go out of our way to associate with the townspeople, join their activities, and share their community interests.[6]

Soon, a mixture of southern hospitality and German courtesy began breaking down barriers, and Huntsville citizens started throwing open doors and welcoming the newcomers to their community. German missile experts were invited to join Huntsville civic groups, and they became enthusiastic participants in community improvement projects.

German *Hausfraus'* pastries were displayed at church bazaars, and church suppers served *Apfelkuchen* (an apple cake) and *Sauerbraten* (German pot roast).

Huntsville youngsters, who were too young to harbor prejudices from the war, immediately accepted their German counterparts in the schools.

Attracting the most attention among the German contingent was Wernher von Braun, who was a rare breed—a scientist with charisma. To those meeting him for the first time, he gave two immediate impressions: that he was moved by a driving force, and that he had been accustomed since birth to being treated as a man of destiny. He also was a man of simple tastes. Each morning, von Braun drove to work in a weather-beaten Chevrolet and parked it in the reserved area at Redstone Arsenal. Swinging the brown briefcase that carried his lunch, he strode toward his office with the buoyant step that had always been his trademark. Guards snapped to attention and were greeted by name. Most replied, "Good morning, Professor!"

Entering the building, he moved down a long corridor to his office, inside of which was a picture of a Chinese, whose message never failed to amuse von Braun. It bore the inscription "No talkee, no tellee, no catchee hellee." Now 39 years old with gray hair at his temples, von Braun continued to put in vigorous 12- to 14-hour workdays. His government pay would never exceed $10,000 annually.

After handling countless administrative details at his cluttered desk all morning, von Braun reached into his briefcase for his lunch, which usually consisted of a bunch of grapes, a peach, or a pear, which he liked to munch while catching up on his technical reading.

Hardly had the Peenemünde vanguard and their families settled in Huntsville than another global crisis erupted—only five years after the World War II carnage ended. On the morning of June 25, 1950, North Korean Communist troops, armed with Soviet-built weapons and tanks, struck southward across the dividing 38th Parallel and invaded South Korea. When a United Nations security council demand that North Korea withdraw its forces was ignored, President Harry Truman ordered American troops to go to the aid of beleaguered South Korea. For the third time in the 20th century, Uncle Sam was sending his sons into battle woefully unprepared and ill-equipped for the violent task at hand.

The war halfway around the world injected urgency into Wernher von Braun's first assignment at Huntsville: developing a 500-mile surface-to-surface missile, an enlarged and more sophisticated version of the V–2. After first being named Ursa and then Major, the 69-foot, six-inch rocket was baptized Redstone, after the arsenal.

While much of the Truman administration's focus was riveted on Korea, Secretary of the Interior Oscar L. Chapman had become alarmed over confidential reports from one of his top scientists who had just returned from Europe. This scientist told Chapman that he had uncovered considerable evidence that the Russians were enjoying great success in luring increasing numbers of German scientists and engineers to the Soviet Union.

On December 21, 1950, Secretary Chapman sent a memo to President Truman, detailing the evidence the Interior Department scientist had brought back from Europe. "His suggestion that we undertake a renewed program for bringing to this country additional scientific talent from Germany seems to be worthy of thoughtful consideration," Chapman concluded.[7]

Harry Truman was impressed. In a "Dear Oscar" reply, the president said: "It seems to me that you have something. Those Russians have been taking over the German scientists and I think we ought to be careful that they don't get a corner on the market. I wish you would discuss the matter with the Central Intelligence Agency [CIA] and then after New Year's we will decide what we ought to do."[8]

Even while building weapons of destruction, Wernher von Braun's vision remained focused on manned space flights. In February 1952, he delivered before a group of American scientists a hard-hitting speech in which he called for a massive U.S. space program to prevent the Soviets from attaining technological superiority:

> With presently available propellants it is possible to build a multi-stage rocket capable of carrying a crew and a substantial amount of payload into a satellite orbit around the earth. [However], the development of a satellite vehicle . . . cannot possibly materialize as a mere by-product of present development programs in the field of guided missiles. A coordinated space program is urgently necessary.
>
> Such a program may be started on a modest scale, but ultimately the conquest of space superiority will succeed only if an effort is made on the grand scale.[9]

Many times, von Braun stuck out his neck and risked his reputation by making space-flight predictions. In March 1952, *Collier's*, a popular, large-circulation magazine, carried a spread, complete with dazzling illustrations, on von Braun's vision of the future in space. "Within the next ten to 15 years," he declared in the article, "the earth can have a new companion, a man-made satellite, one inhabited by human beings and visible from the ground as a sedately moving star."[10]

Wide media exposure triggered a barrage of criticism aimed at von Braun. A science writer for the *Washington Post* barked: "This man has no business criticizing the Administration or anyone else for the nation's missile and space lag."

"Who does von Braun think he is—the Christopher Columbus of space?" asked an editorial in the *St. Louis Star-Times*.

Many high priests in American science scoffed that space exploration was not worthy of a major national effort, and a large number of American universities became "fort-diggers" in bitter opposition to space exploration.

Even a few of the Peenemünde men frowned on von Braun's public space-travel predictions. One of them, Paul von Schroeder, a mathematician, accused von Braun of being an opportunist, one who grabbed credit that other men should have had. What was more, Schroeder exclaimed, there was too much of the huckster in von Braun. To that charge, von Braun replied: "I admit to being a two-headed monster—part scientist, part public relations man."

A columnist in *The New Yorker* was far more sympathetic: "Wernher von Braun is regarded by the conventional rocket men with the mixture of suspicion and admiration that must have been felt by cosily established clerics toward Saint Francis of Assisi or Peter the Hermit. He worries and frightens them with his technological vision. When he talks to the lay public about his confident plans for voyaging into space, they accuse him of 'preaching to the birds.' "

Meanwhile, the Korean War grew more intense and threatened to ignite World War III, so the Army's requirements changed, and the desired range of the Redstone rocket was reduced from 500 miles to 200 miles. This would permit the missile to carry a nuclear warhead or be a mobile weapon capable of being launched on battlefields by combat troops.

Meanwhile, a steady stream of reporters from around the nation descended on Huntsville to send back stories on activities at Redstone Arsenal. Like all scribes, each was eager to get a scoop and seemed to be disappointed that the Peenemünde Germans had not performed miracles.

Few, if any, of the newsmen were aware that there were no guidebooks to help solve perplexing problems. Rather, the rocket had to be designed, built, and tested. Always the initial test disclosed a need for alterations, so back to the drawing board went the scientists and engineers to redesign the Redstone and have new parts manufactured. This tedious and time-consuming process had to be repeated over and over.

One day, a reporter bluntly demanded to know why there had not been more results. Keeping his composure as always, von Braun pleasantly replied: "My dear fellow, do you know that a large guided-missile system involves from sixty to eighty thousand engineering changes on the missile alone before its first flight?"

Wernher von Braun was confronted steadily by another kind of vexing problem—keeping the Peenemünde team intact. Big industry was trying to entice his experts by dangling large paychecks before their eyes. Nearly all of the original team members, however, remained with the Army missile program. One associate said of von Braun: "Working for the government you don't wind up a rich man. But working for him, you feel you're a member of a great team."[11]

Von Braun himself had a standing offer from a major corporation: a vice president's post at an initial salary five times the size of the one he was drawing at Redstone Arsenal—plus bonuses. He was sorely tempted. A second daughter, Margrit Cecile, had been born to the couple, and he had to consider his family's future financial security. However, von Braun rejected the vice presidency and the fat salary out of loyalty to the U.S. Army.

On August 20, 1953, the Redstone was test-fired successfully. So a few months later, when the Space-Flight Committee of the American Rocket Society (ARS) proposed that a satellite be boosted into orbit, Wernher von Braun took his case directly to Washington. On June 25, 1954, he met with top military officials in Room 1803, T–3 Building, of the Office of Naval Research.

There were several American high-altitude research projects in progress, but no satellite program, so von Braun presented a plan to hurl a five-pound satellite into orbit with a souped-up Redstone. The man-made moon could be launched into orbit 200 miles above the earth without a major new alteration in the Redstone, and a heavier satellite, weighing 15 pounds, could follow shortly afterward.

Rear Admiral Frederick R. Furth, chief of Naval Research, gave the green light to the concept, code-named Project Orbiter. It would be a joint Army-Navy operation. Designing, developing, building, and launching the souped-up Redstone missile would be the task of the Army group at Huntsville, while the Navy would design and build the satellite.

Countless meetings were held between key Army and Navy personnel, and September 1956 was targeted as the date for launching the world's first man-made satellite into orbit. None of the eager scientists and engineers could know that faulty reasoning and an overall lack of understanding at high levels of government would scuttle America's golden opportunity to put history's first man-made satellite into orbit ahead of the Russians.

In the meantime, rocket launchings were continuing at Cape Canaveral. Rocco A. Petrone, who years later would hold the crucial job of director of launch operations at the Cape, remembered:

> In the early days, there had been a lot of hit-and-miss, seat-of-the-pants business in launching rockets—and, insofar as the general public was concerned, a lot of black magic.
>
> Some odd things happened. I recall a Redstone launch in May 1954, when I was still in the Army, that we had to delay because someone reported a fishing skiff offshore, a dangerous place to be at lift-off time.
>
> We sent out guards to warn the fishermen to get out of the way. By the time the guards reached the spot they found the boat was up on the beach and the fishermen were gone. Then we had to send someone to find the guards and tell *them* to come back, and that took another hour before we could launch.
>
> At other times the countdown would be held up because diesel engines powering a tractor didn't work, or a key to the tractor had been lost.
>
> But we learned as we traveled this long road.

April 1955 marked the fifth anniversary of the arrival in Huntsville of the Peenemünde team and their families. Over that half-decade, the Germans had become deeply integrated into the community. Initially strongly opposed to the creation of a missile center in their midst, the Chamber of Commerce had adopted a new slogan for Huntsville—Rocket City, USA.

Walter Wiesman, a management expert, was elected president of the Huntsville Junior Chamber of Commerce while still a resident alien and later would become state president. He also would be named chairman of the Huntsville Community

Association, vice president of the Civic Council, and publicity chairman of the county Red Cross.[12]

Hannes Luehrsen, once the chief architect at Peenemünde and now head of Redstone Arsenal's planning branch, laid out a belt highway around Huntsville to relieve the heavy traffic congestion that was threatening to strangle the rapidly mushrooming city. His plan was opposed bitterly by downtown merchants, who felt that their sales would drop off drastically. But most of the citizens were boosters of the belt highway project and it was approved by the Huntsville planning commission, helping to transform Rocket City into a booming metropolis of some 65,000 people.[13]

Ernst Stühlinger was among the Germans who founded the Rocket City Astronomical Association. Its members were responsible for helping to make Huntsville one of the few cities of its size and to have its own observatory—a boon to the high school system's science departments. Collaboration between citizens and the Redstone scientists resulted in founding of the Huntsville Civic Orchestra, with German names especially prominent among its 60 musicians.

On April 15, 1955, nearly 1,300 Huntsville residents shoehorned their way into the Huntsville High School auditorium to witness Wernher von Braun and 39 members of the Peenemünde team, along with wives and children, become citizens of the United States. Earlier, Walter Thiel, Walter Schwidetzky, and Theodore Buchold had become citizens by marrying American women.

Mayor R. B. "Specs" Searcy, who had proclaimed April 15 to be New Citizens Day in Huntsville, said: "I am glad you have chosen us. I know of no group we have enjoyed more in our community."[14] Von Braun responded in an emotional tone: "This is one of the proudest and most significant days of my life, and I know I am speaking for all of us." The Huntsville citizens shook the rafters with cheers.[15]

Late that night at home, von Braun flipped on the radio and heard the news that dampened the warm afterglow of the citizenship ceremony of the day. A newscaster said that the Kremlin had revealed that its physicists were finishing a flying laboratory to study rays that were held from reaching the earth by dense layers of tiny particles in space. This craft, the Soviet spokesman had boasted, would circle the earth as a satellite and "solve the problem of interplanetary travel."

Von Braun held no doubt that the Soviets had made gigantic technological breakthroughs in their all-out secret effort to reach space ahead of the Americans. These advances, he was convinced, included the development of rockets with far greater thrust than U.S. missiles had.

During the past two years, the Army group at Redstone Arsenal had formulated several proposals for developing a much larger rocket with far more thrust than anything known in the world to date. But each time, the proposals were shot down by Secretary of Defense Charles E. Wilson.

Since resigning as chairman of General Motors Corporation to accept the appointment by President Eisenhower in 1953, Wilson had a reputation in Washington of being brutally blunt in his dealings with others. A hulking man with a thick shock of white hair, Wilson had raised the ire of senators at his confirmation hearings by insisting on addressing them condescendingly as "you men." Tabbed "Engine Charlie" by the not too adoring Washington press corps (because of his earlier connection with General Motors), Wilson gruffly demanded each time a new-rocket plan was submitted to him by the Army, "Show me an immediate military need for this object."

It was an impossible demand. Had there been "an immediate military need" it would have been already too late to design, test, and build the new missile.[16]

In the meantime, barely concealed behind the scenes, the Navy and the Air Force were scrambling to boost the world's first man-made satellite into orbit. Although tentatively involved in a joint venture with the Army in Orbiter, the Navy proposed using its version of the V–2, called the Viking, while the Air Force countered by proposing its undeveloped Atlas.

Media columnists and editorial writers leaped into the burning dispute, each one having his favorite service to boost. "Space craze" was sweeping the nation and the man on the street (and in the bars) began arguing vehemently over the capabilities of the Redstone, the Atlas, and the Viking.

All the while, the space mills had been grinding outside of the U.S. military establishment. The National Science Foundation recommended that the satellite project become a part of the American contribution to the International Geophysical Year (IGY). It was recommended that it would be far more appropriate for the United States to boost a satellite into orbit by a means other than weaponry.[17]

Scheduled to start on July 1, 1957, the IGY was to continue for 18 months. Its purpose was for 40 nations, including the Soviet Union, to pool their technological know-how in the exploration of earth and its atmosphere. Fields to be explored including meteorology, longitude and latitude determinations, geomagnetism, gravity measurements, solar activity, cosmic rays, and oceanography.

On July 29, 1955, Washington reporters were alerted that the White House would have a "highly important announcement" to make at 1:30 P.M. Despite the atmosphere of secrecy, the Washington news faucets already had been leaking. From both the Pentagon and the Capitol came rumors that the news concerned the satellite program.

At the appointed hour, about 100 newsmen, pencils and pads at the ready, thronged the White House conference room, a rare stage for such an event. Presidential Press Secretary James C. Hagerty, himself a former newsman, was flanked by a battery of scientists: Alan T. Waterman, director of the National Science Foundation; J. Wallace Joyce, head of the foundation's Office for the IGY; S. Douglas Cornell, executive officer of the National Academy of Sciences; Alan H. Shapley, vice chairman of the National Committee for the Geophysical Year; and Athelstan F. Spilhaus, a member of that committee's executive group.

Jim Hagerty announced that Dwight Eisenhower had given his approval to the launching of a satellite as part of the U.S. contribution to the IGY. The operation would be known as Project Vanguard. An entirely new, nonmilitary rocket would be built for this purpose by the Naval Research Laboratory, under the direction of the National Academy of Sciences and the National Science Foundation.

Project Vanguard would launch history's first man-made moon, a globular object about the size of a basketball, it was explained. The satellite was to flash around the earth about once every 90 minutes at a speed of 18,000 miles per hour in a fixed path 200 to 300 miles above the ground.

Project Orbiter has been shoved into limbo.

At Huntsville, Wernher von Braun was bitter about the news from Washington and predicted, accurately, that it would go down in American history as a fateful "wrong decision" that would assure the Russians first reaching space. It particularly galled him that the National Science Foundation and the National Academy of Sciences had failed to consider that the Vanguard was based on rocket hardware that did not exist, whereas the Redstone was constructed with tested parts. Based on his 15 years of experience with the V–2, von Braun knew that a proposed rocket, although perfect on the drawing board, was far more likely to result in an explosion than in a celebration.

Spokesmen in the White House and the experts involved touted Vanguard as being "dignified," as it would not be a weapon. "I'm all for dignity," von Braun told confidants. "But this is a Cold War tool. How dignified will the United States be if a man-made moon of Soviet origin suddenly appears in the sky?"

# 18

# Billy Mitchell of the Space Age

Across the land, the "big satellite story" from Washington touched off a torrent of media hype. Sparked by White House press agentry, the likes of which would have gladdened the hearts of Madison Avenue hucksters, radio and television networks interrupted regular programs with bulletins of America's looming space spectacular. Newspapers splashed stories and photographs across front pages. Seldom since World War II concluded had such a barrage of media publicity flooded the nation.

Widely regarded as America's most influential newspaper, the customarily staid *New York Times* plunged overboard. A blaring headline in its December 7, 1955, issue shouted: "U.S. TO LAUNCH EARTH SATELLITE 200–300 MILES INTO OUTER SPACE." Nearly the entire front page of the *Times* was devoted to Project Vanguard, as were three inside pages. The *New York World-Telegram and Sun* hailed Vanguard as "the greatest adventure of modern times."[1]

The big story was played as though the event had already taken place and had been a whopping success. Print editors and broadcast news directors, perhaps dazzled by the revolutionary nature of the event, overlooked a crucial point: a man-made moon in orbit was only a theoretical proposition.

In Washington, most members of Congress were skeptical. Senator Richard B. Russell, Democrat of Georgia and chairman of the Senate Armed Forces Committee, expressed "grave doubts about the wisdom of giving the Communist world the benefit of expenditures and effort that we are putting into this project."[2] Congressman Carl T. Durham, Democrat of North Carolina, declared that he opposed "swapping information with the Russians on anything."[3]

Late in 1955, many in official Washington were alarmed when the Soviet newspaper *Vechernaya Moskva* (Evening Moscow) quoted an unidentified scientist: "Our rocket engines show that the necessary speeds and thrusts can be attained, and it is we, the Communists, who will engage soon in real interplanetary flight."[4]

With the arrival of 1956, Democratic leaders were searching for a compelling campaign issue to hurl against Dwight Eisenhower in the presidential election in the fall. They settled on what they would brand as the Missile Gap, meaning the Soviets' alleged superiority over the United States in space development.

Leading the Democratic charge was 54-year-old Stuart Symington, a tall, silver-haired man with classic facial features who was a senator from Missouri. Symington himself, it was reported, had presidential ambitions. Since Symington, a native of Amherst, Massachusetts, had been the nation's first secretary of the Air Force back in 1949, his viewpoint on missiles held credence with many.[5]

Symington, along with Henry "Scoop" Jackson of Washington state, also a presidential hopeful, began pounding that the Russians had developed an IRBM (intermediate-range ballistic missile) with a range of 1,500 miles and demanded to know what Eisenhower planned to do to "close the gap."

At a meeting of Republican leaders in the White House, Joseph W. Martin, Jr., who had entered Congress from Massachusetts in 1925 and was now House GOP (Grand Old Party) floor leader, told Eisenhower that the Democrats were charging that the president "doesn't know what's going on." Eisenhower chuckled and replied: "When I listen to Symington, I think I don't!" Turning serious, the president declared: "I want guided missiles, as soon as possible."[6]

At a news conference in the White House, a reporter asked Eisenhower to comment on Senator Symington's charge that the United States was lagging far behind the Soviet Union in the development and production of guided missiles. Chewing on the earpiece of his horned-rim glasses as was his habit in moments of deep contemplation, the president replied sarcastically, "Well, I am always astonished at the amount of information that others get that I don't get."[7]

Eisenhower told the reporters that the $1.2 billion that he had put in the 1956 budget for the missile program more than doubled the half-billion figure for 1955. "Now the Democrats are critical of U.S. missile progress, and at the same time are demanding to know what we are doing with all that money," Eisenhower declared.

As the weeks passed and the 1956 presidential sweepstakes began to heat up, Eisenhower became irritated at the relentless pounding by Senators Symington and Jackson and other Democratic bigwigs over the Missile Gap. Chain-smoking Eisenhower, who, when angered, could swear in a style that would make a mule skinner blush, told his confidants in earthy terms: "There's just so much money you can throw at something. There's just so many scientists."

In a carefully worded announcement designed to indicate that he was not reacting to the barrage of criticism launched by the Democrats, the president said that, on the recommendation of scientists, the United States would expand the missile program. Although concerned about the waste involved because of duplication of effort, Eisenhower decided to split the program. The Army (Jupiter) and the Navy (Thor) would develop intermediate-range ballistic missiles, while the Air Force would have two separate intercontinental ballistic missile projects (Atlas and Titan).

Eisenhower said repeatedly that he hoped to "take the missile program out of politics,'' but the Gap charge would not go away. Reporters besieged the president about the Gap. And the venerable Bernard Baruch, a 76-year-old financier who had been an unpaid advisor to every president since Woodrow Wilson in 1917, leaped into the fray.

On his own volition, Baruch sent Eisenhower a series of alarming letters in which he stressed the critical need for accelerated progress in missile development and urgently recommended that huge additional funds be poured into the program. Although the Eisenhower White House regarded Baruch as a man puffed up with his own importance and dealing with him a bore, the president decided to invite him to a private meeting ''because of his standing and reputation in the public mind.''

Eisenhower spent two hours on the afternoon of March 28, 1956, with Baruch. Later the president wrote in his diary: ''I tried to show him that we are already employing so many of the nation's scientists and research facilities that even the expenditure of a vastly greater amount could scarcely produce any additional results.''[8]

Meanwhile, for a project only in the development stage, Vanguard continued to be flooded with media publicity to an unprecedented degree. Concerned by the enormous disillusionment that would strike the American people should Vanguard fail to meet its lofty goal, John P. Hagen, director of the operation, issued a public statement. Hagen, a 47-year-old physicist, emphasized that there might be but one chance in six that the initial launching to put a man-made moon into orbit would succeed. His warning was brushed off by the media as the ''usual protective official hedging.''

Hagen's function was to coordinate the scores of scientists and hundreds of private companies involved in the crash project. Born in Amherst, Nova Scotia, Canada, Hagen came to the United States in his teens and later received bachelor's and master's degrees from Boston University and a Ph.D. in physics at Georgetown University in Washington, D.C.

Wearing gold-rimmed glasses and usually smoking a pipe, Hagen was an unassuming, soft-spoken yet decisive man with a dry sense of humor. That he became a scientist at all was purely by chance, for he aspired to be a lawyer. ''In my senior year,'' Hagen explained to a friend, ''I was offered a teaching fellowship in physics at Connecticut's Wesleyan University. That paid money; law school didn't. So I became a scientist.''[9]

As the weeks passed, Project Vanguard was beset by monumental problems that deeply frustrated those directly involved. Work was done in a fish bowl, with much of the world peeking in. Generated by a flood of hype pouring out of Press Secretary Jim Hagerty's office in the White House, print and broadcast media reported profusely on each minor problem and design change.

The pressure was enormous on Hagen; his youthful deputy, J. Paul Walsh; and other top Vanguard officials. The timetable was too tight. The initial funds had been too small, and those doling out the dollars in Washington engaged in parsimonious feuding with one another. Then there was the biggest time-delayer of all: a policy set by Washington bureaucrats that decreed, "in the spirit of democracy," Vanguard must have the "full participation of industry."

Such a decree would be logical and necessary for production, but it proved to be a huge handicap when experimental work was being done on a crash project. If Vanguard had a need for a few instruments or switches, for example, the work could have been done rapidly by a government research laboratory, staffed by scientists and equipped with sophisticated machinery. However, that laboratory could not be utilized; instead, competitive bids had to be let.

This tedious and time-consuming procedure meant that, in order to obtain the handful of instruments and switches, 40 contracts, in quintuplicate, together with hundreds of pages of related correspondence, had to be drawn up. A blizzard of telephone calls, memos, and telegrams then would fly back and forth, and months would pass before the instruments and switches arrived at the designated Vanguard work site.

Although Project Orbiter had been relegated to the ash can, William Pickering and Jack Froehlich, both prominent American scientists who had long been conferring unofficially with Wernher von Braun, flew to Huntsville covertly to explore means for resuscitating the Redstone missile as a possible backup should Vanguard fail. Under the veil of night, Pickering, Froehlich, and von Braun met in a secret session at the home of Colonel Miles Chatfield. Discussions continued almost until dawn.

Perhaps as a result of that conference, Wernher von Braun used his ample powers of persuasion to coax from the Pentagon permission and funds to build a few more Redstones, explaining that he wanted to iron out problems of reentry into the atmosphere.

Since even the mention of Project Orbiter was *verboten* in the Pentagon, the Army called its new design the Jupiter C. It would be a three-stage, souped-up Redstone with clusters of Sergeant rockets in its nose. Jupiter C would have the capability of accommodating, with only a minor alteration, a small fourth-stage rocket on top of the Sergeants, allowing it to place a fully instrumented, 30-pound satellite into orbit.[10]

Late on the night of September 19, 1956, Wernher von Braun was looking with pride on the Jupiter C as it was bathed with floodlights on its launchpad at Cape Canaveral. Its initial test-firing would take place after dawn. Jupiter C was crowned with a satellite dummy going along for the ride.

Only hours before blast-off, Pentagon officials in Washington had grown nervous: the Army might "mistakenly" boost a man-made moon into orbit ahead of Vanguard, then apologize for the "error" later. Around midnight, von Braun received an urgent telephone call from Major General John B. Medaris, the new commander at Redstone Arsenal, which had a new name, the Army Ballistic Missile Agency (ABMA). Medaris ordered von Braun personally to inspect the satellite dummy to make certain that it was not "live" and capable of being put into orbit. Under heavy pressure from the Pentagon, Medaris declared, "What we don't need is an 'accidental satellite'!"

A few hours later, Jupiter C soared into space. Flawlessly, it flew for 3,300 miles down the Atlantic missile range, reaching a speed of 16,000 miles per hour and a height of 680 miles—faster and farther than an American missile had ever traveled.

Despite the resounding success, von Braun and others on the Army team were subdued. They knew that a golden opportunity had been lost to put a satellite into orbit ahead of the Russians.

For weeks, the Pentagon blacked out any mention of the astonishing feat. But news faucets have a curious penchant for leaking at crucial points, so a month after the flight *Life* magazine hit the streets with a dramatic picture spread on the Jupiter C spectacular. But even *Life* did not know that the dummy man-made moon could have been put into orbit had it been live.

Seven more Jupiter C missiles now were available at Huntsville, and the Army group waited eagerly for word from Washington to put a satellite into orbit. The order never came. What did arrive, on November 26, 1956, was a directive from Secretary of Defense Charles Wilson—a stunning blow to the Army missile program.

Acting to end what was described as a long-standing dispute between the Army and Air Force, Secretary Wilson ruled that the Air Force would take control of all missiles with a range of more than 200 miles, and the Army would confine its research and development to rockets with a range of no more than 200 miles.

The "clarification" of the services' roles by Wilson was the most sweeping since the definition of missions by Secretary of Defense James Forrestal in 1948. His action, Wilson told reporters, was mandatory to avoid "unnecessary duplication and functions," conceding that his decision was a "resolution of divided opinions within the Joint Chiefs of Staff."

Civic and government leaders in Huntsville were shocked by Secretary Wilson's edict, which Mayor "Specs" Searcy and others interpreted as a first step in closing down Redstone Arsenal, where 6,000 civilians were employed. At the arsenal, General Medaris called his staff together and gave strict orders not to challenge the Wilson decision or to take any other part in the public controversy that had erupted.

Medaris reminded his aides that the decision merely gave operational control of the Jupiter C to the Air Force and that it specifically mentioned that the Army was not prohibited from making "limited feasibility studies of IRBMs."

Privately, countless Army officers believed that the Pentagon edict sounded the death knell for that service's role in missile development. One who decided to take covert action to overturn the ruling was 41-year-old Colonel John C. Nickerson, Jr., a West Pointer, decorated combat leader in World War II, and now field coordinator for the Army Ballistic Missile Agency at Huntsville.

Nickerson, a native of Paris, Kentucky, had reached an anguished decision, one that might end his Army career or even send him to the penitentiary. He would become the ''Billy Mitchell of the Space Age.'' Thirty-one years earlier, in 1925, Brigadier General William Mitchell became one of the most controversial figures in American military history when he was court-martialed for defiance of his superiors.

Billy Mitchell had been air advisor to General John J. Pershing, commander of U.S. forces in Europe in World War I, and led large air units in combat. After the conflict, Mitchell became assistant chief of the Air Service and leading advocate of air power. Because airplanes were then limited in size and range, many hidebound generals in the older services branded Mitchell as a troublemaker and extremist.

Mitchell took his plea for air power to the public through books, magazine articles, newspaper interviews, and speeches. Superiors warned him to keep his mouth shut, but he refused to do so. When he proved the potential of air power by sinking three warships in an air-attack demonstration, much of the nation was won over. Stung by Mitchell's feat, however, the Army ordered him court-martialed, found him guilty of insubordination, and ordered him suspended for five years. Rather than accept the suspension, 45-year-old General Mitchell resigned.

Now, three decades later, Colonel John Nickerson prepared a document called ''Considerations on the Wilson Memorandum,'' which took sharp issue with Secretary of Defense Charles Wilson. Nickerson, a six-foot-tall, rawboned, bespectacled man, asserted the superiority of the Army's Jupiter C and insisted on the Army's crucial need for operating use of its own intermediate-range ballistic missiles.

In mid-December 1956, Nickerson circulated his lengthy memorandum, together with supporting documents, among influential individuals he thought might persuade Secretary Wilson to modify his controversial decision. On the colonel's distribution list were Erik Bergaust, managing editor of *Missiles and Rockets*, a magazine; Drew Pearson, syndicated columnist; William F. Hunt, chairman of Reynolds Metals Company; John A. Baumann, chief executive officer of Radio Corporation of America; and several members of the Alabama delegation in Congress.

Washington-based Drew Pearson's legman was sent scurrying to the Pentagon to see if he could stir up a rebuttal from the Air Force. But the Air Force refused to rise to the bait and notified the Army, which ordered the Pearson copy of the Nickerson memorandum confiscated. The Pentagon was thrown into what one Army major there described as a ''tizzy.'' Secretary of the Army Wilber Brucker began padding around Capitol Hill in person picking up other copies from Alabama congressmen.

Back at the 40,000-acre Redstone Arsenal, Army security investigators barged into the antebellum (1817) home of Colonel Nickerson, his wife, Carol, and their four children, searched it from attic to basement, and refused to let anyone in or out for 24 hours. Confiscated from the Nickerson residence were several secret documents, which, the Army charged, were lying in the open, instead of being kept under lock and key as required.

In January 1957, multiple charges were lodged by the Army against the colonel and he was ordered to be confined to the Redstone Arsenal premises pending his court-martial. What came to be known as "the Nickerson case" was on.

Colonel Nickerson began busily getting in step for the trial with heavy speculation among the ranks of the U.S. military that the Pentagon might have on its hands a new Billy Mitchell. Nickerson's attorney, Robert K. Bell, former law partner of Alabama's Senator John Sparkman, hinted to the throng of reporters covering the affair that he might well grill high defense officials, including Charles Wilson.

When the court-martial was convened at Redstone Arsenal on June 26, 1957, Colonel Nickerson faced serious charges of espionage (spying), perjury (for making false statements to Army investigators), moral turpitude, and failure to safeguard secret information.

Ernst Stühlinger, a Peenemünde team member, testified that some members of his group had been working together for 20 years and had more missile experience than any other group in the world. If the Wilson decision kept it from continuing its work, he said, the interests of the Army and the United States would suffer. This was especially true, Stühlinger added, in view of great missile progress known to have been made by the Soviet Union.

Within an hour and a half of the opening of the trial, Lieutenant Colonel William G. Barry, the chief prosecutor, announced the withdrawal of the espionage, perjury, and moral turpitude charges. Then, standing erect, his decoration ribbons blazoned on his chest, John Nickerson pleaded guilty to the remaining charge of violating security regulations.

General John Medaris, Nickerson's boss, took the stand as the trial continued in order for the court to decide on a sentence for the defendant. Medaris testified that the colonel had "absolutely and diametrically" violated his (Medaris') instructions not to challenge the Wilson decision. "Colonel Nickerson has violated the fundamental military code, and does not have any further potential value to the Army," Medaris declared.

Under questioning by Ray H. Jenkins, one of the defense lawyers who had gained fame in 1954 as counsel for the Senate committee in the heated Senator Joseph McCarthy hearings, General Medaris conceded that Nickerson's motives were sincere, that he had "a good end in mind," and that the Jupiter C was superior to the Air Force's Thor.

On June 29, the court, consisting of five generals and five colonels, spent 40 minutes in secret session deciding on John Nickerson's sentence. Major General Crump Garvin, president of the court, read the verdict: the war hero and missile expert would receive an official reprimand, be suspended from rank for one year, and forfeit $100 per month pay for 15 months (out of his $900 monthly pay, quarters, and subsistence). The maximum sentence could have been 30 years in prison and dismissal from the Army.[11]

Four weeks later, the Pentagon announced that Colonel Nickerson was being assigned to an obscure job in the Panama Canal Zone. Like a policeman being sent to a beat in the sticks, Nickerson would be in charge of engineering at the Canal Zone's Fort Clayton. A Pentagon spokesman stressed that he would not be "associated with guided missiles or research and development in any way."[12]

While Project Vanguard was moving forward by fits and jerks, thanks to the perseverance and skill of John Hagen and his associates, General John Medaris played host to Neil McElroy, who had been named by President Eisenhower to replace "Engine Charlie" Wilson, who had resigned the post. McElroy, the former chief executive of the Procter & Gamble Company, and a bevy of high-ranking Pentagon officers arrived on October 4 for a get-acquainted tour of Redstone Arsenal.

With the defense secretary-designate were Secretary of the Army Wilbur Brucker, Army Chief of Staff Lyman Lemnitzer, and Lieutenant General James M. Gavin, who had gained fame in World War II as the 37-year-old leader of the crack 82nd Airborne Division and the youngest two-star general since the Civil War. Gavin now held a Pentagon post as chief of research and development for the Army.

That night, a dinner on the post had as guests the Pentagon contingent, key officials of the Army Ballistic Missile Agency, and government and civil leaders in Huntsville. General Medaris and Wernher von Braun looked on the occasion as a golden opportunity to make a sales pitch to the new secretary of defense for an expanded missile role for the Redstone group. They planned to pull no punches.

While the group was waiting to take seats, Neil McElroy was talking shop with General Medaris and von Braun. Suddenly, an excited public relations officer for the ABMA bustled up breathlessly to the three men and blurted: "The Russians have just sent a satellite into orbit!"

Secretary McElroy, who had not yet gotten his feet wet in the mysterious machinations of international politics, looked stunned. How could this have happened? Weren't the Americans supposed to be ahead in the race for space?

Von Braun, not surprised over the Soviet feat, turned to McElroy and words expressing his pent-up emotions and frustrations tumbled out: "We knew that the Russians were going to do it! We have the hardware on the shelf [Jupiter C]. For God's sake, turn us loose and let us do something. We can put up a satellite in 60 days, Mr. McElroy! Just give us the green light!"[13]

"Make that 90 days," General Medaris interjected.

It would turn out to be 80 days.

# 19

# "Goldstone Has the Bird!"

Early on the morning of October 5, 1957, a grim President Dwight Eisenhower was presiding at an urgent conference at the White House. Among those present were Department of Defense officials and assorted scientific advisors. The topic: How had the Russians managed to win the space race?

Like shocked Americans from coast to coast, Eisenhower learned that the Soviet moon was named *Sputnik* (traveling companion), which was 23 inches in diameter and orbited the earth once every hour and 35 minutes at a maximum altitude of 560 miles.

*Sputnik* scored an enormous propaganda coup in the Cold War struggle. Previously, most of the world—including President Eisenhower—presumed that the United States was the leader in applied science and technology.

Now, at the solemn White House conference, Eisenhower was told that two unidentified Army officers had stated the previous night that the Army's Jupiter C could have placed a satellite in orbit many months ago but that the administration had put all its eggs in one basket—Project Vanguard.

Eisenhower looked startled. "Is this true?" he asked Deputy Secretary of Defense Donald Quarles. It was worse than true, Quarles replied candidly: a Redstone could have put up a man-made moon two years earlier.

Lame-duck Defense Secretary Charles Wilson, whose impending departure from government already had been announced, airily brushed off the Sputnik as a "neat scientific trick." But the president declared, with considerable intensity, that "we might have to go to a 'Manhattan Project' type approach in order to get forward in this matter."[1]

Eisenhower then instructed Wilson to rescind the cost-cutting restrictions on overtime work at missile research installations and to "get the Redstone people into the business of putting a satellite into orbit as soon as possible."[2]

On November 2, 1957, the Soviets again grabbed the global spotlight by launching *Sputnik II*. It carried a mongrel dog named Laika and a payload of 1,100 pounds. Laika's heartbeat was telemetered: it had grown more rapid during the ascent but settled back to normal within minutes after the satellite went into orbit.

Uncle Sam had egg on his whiskers. A gleeful broadcaster on Radio Moscow declared: "Our space dog is now flying over Washington where senators are worried about whether it will bite." Yet another Soviet barb: "The trouble with the American satellite is that they can't find a dog small enough to ride in it."

After whirling the earth for 100 hours, *Sputnik*'s batteries ran down and Laika was painlessly put to death in outer space.

Meanwhile, time was drawing near for the Vanguard launching. On December 5, 1957, the eve of what the press was calling "America's answer to the Sputniks," headlines from coast to coast hailed the imminent space spectacular. Trumpeted a banner in the *Pittsburgh Sun-Telegraph*: "U.S. MOON—MINUTES TO GO."

With its customary alacrity and foresight, the Associated Press flashed an advance story to hundreds of client newspapers and broadcast stations, to be set in print or written into stories for release to the public moments after the next day's launching. It began: "The radio-signaling baby moon circling the earth is the U.S.'s reply to Russia that it too can stake a claim to the space frontier."

For miles around Cape Canaveral on the following morning, schoolchildren, housewives, workers, and members of the armed forces poured into streets, yards, roadsides, and public beaches. Down Highway A1A, thousands of tourists, drawn from all parts of the nation and many foreign countries, moved outside from such luxury motels as the Sea Missile, the Starlite, and the Vanguard.

In a spider-web gantry on its launchpad stood Navy Test Vehicle 3(T–3), a tall, three-stage rocket, the sun sparkling off a thin coat of frost crystals on its silver and jet-black skin. T–3 (known to the public as the Vanguard) was to gain an altitude of 300 miles and a speed of 8,500 miles per hour. The third stage, through a solid-fuel propellant, would provide the additional 50 percent of the orbital velocity of 18,000 miles an hour to put into orbit a grapefruit-sized satellite 6.4 inches in diameter.

At 8:45 A.M., the big red ball that signified an impending launching went up. It showed clearly against the soft blue sky streaked with occasional puffy clouds. A few minutes after 11:00 A.M., a white trail of liquid oxygen seeped from the Vanguard and vaporized off to the west. Experienced bird-watchers knew that this meant firing was near. By now, the throngs numbered in the hundreds of thousands, and they were in a Mardi Gras mood. "It's kind of like being on hand to watch Christopher Columbus come ashore!" an excited man shouted to a friend.

At his command post, J. Paul Walsh, the deputy director of Project Vanguard, was talking on a special telephone circuit to his boss, John Hagen, in Washington.

Everything was in good condition; the rocket was in good shape, Walsh reported. "For a time we were sweating out the wind, but it came down to an acceptable level," he told Hagen.

As the moments ticked past, an eerie hush of anticipation hovered over the masses in and around the Cape. At 11:44 A.M., the firing button was pushed. There was a thunderous rumble on the launchpad, and the ponderous rocket began to lift: two feet . . . four feet. Then T–3 toppled over and the launchpad ignited in a caldron of flame and smoke. "America's answer to the Sputniks" had turned into a gargantuan disaster—with the entire world looking on.[3]

Gasps erupted from tens of thousands of bird-watchers. They could not believe their eyes. A day later, blaring newspaper headlines from coast to coast poured ridicule and sarcasm on Vanguard:

COLD WAR PEARL HARBOR
—*San Francisco News*

SAMNIK IS KAPUTNIK
—*Washington Daily News*

9–8–7–6–5–4–3–2–1—PFFT
—*Los Angeles Herald Express*

OH, WHAT A FLOPNIK
—*Chicago Sun-Times*

Shock waves swept across the United States. Said a university professor in Boston: "This is our worst humiliation since Custer's Last Stand!" President Eisenhower, at his farm in Gettysburg, Pennsylvania, where he was recuperating from a mild cerebral attack a week earlier, told reporters, "Of course, I'm disappointed." In an interview with radio station WOR in Washington, Senator Clinton P. Anderson, a New Mexico Democrat, said, "It was pretty well known that (the T–3) was not ready but the demand was to shoot it now, no matter what happens."

Senator Hubert H. Humphrey, a Minnesota Democrat with presidential ambitions, harshly condemned President Eisenhower. "I am saddened and humiliated by the cheap and gaudy manner in which the Administration has gone about the business of trying a last-ditch rush to launch a satellite—and failing," Humphrey said in a speech.

Robert C. Truax, president of the American Rocket Society, told a group at the Statler Hotel in Washington: "With the entire nation looking down their throats, I would have been surprised if it had been a success for that reason alone." Republican Congressman Gerald R. Ford of Michigan voiced a similar viewpoint. "The advance publicity buildup put overemphasis on a minor technical defeat," he declared.

Members of the Soviet delegation to the United States in New York could barely conceal their delight over the debacle. Still basking in the glow of two Sputniks, the Russians inquired slyly of American delegates if the United States were interested in receiving aid under the Soviet Union's program of technical assistance to backward nations.

The engineers from General Electric (which had built the T–3 rocket) blamed the Martin Company (which had built the engine). Along with its reputation, the Martin Company was hard hit financially. In the wake of the launchpad caldron, Martin had the most violent reaction of any stock listed on the New York Stock Exchange. A rash of sell orders caused governors of the exchange to suspend trading in Martin stock at 11:52 A.M. on the following day.

In Washington, Assistant Secretary of Defense for Public Affairs Murray Snyder, after blaming subordinates, newsmen, and the beachside location of Cape Canaveral for the stupidities of the Vanguard carnival of failure, resolved—or was told from on high—never to let the media hoopla go as far again.

Ever since launchings began at the Cape, the commotion had been covered from nearby beaches by reporters with high-powered field glasses, photographers, and TV cameramen with long-range lenses. At first, the bird-watchers found it relatively easy to discover when important launchings were scheduled from cultivated sources on the base. But the Vanguard fiasco changed everything. For one thing, newsmen themselves realized their competition for headlines was partly to blame for the heavy hype leading up to the Vanguard explosion. Most reporters did not want that to happen again.

Canaveral also tightened security on strict orders from the Pentagon. Major General Donald N. Yates, the base commander, cracked down on missile workers with loose tongues. Local police barred those with cameras from prime vantage points on the Cape.

After a month of wild rumors, complaints, false alarms, and assorted other disruptions, some 50 newsmen huddled with General Yates to work out a solution. Yates agreed to give advance notice on scheduled launchings if reporters would avoid dramatizing interservice missile rivalry and write no advances on missile launchings. "The whole world is watching us," Yates declared with considerable merit.

Most newspaper editors and broadcast news directors okayed the shaky compromise. But as the weeks went past, some newsmen began to feel that they had gotten the short end of the deal.

A chilling night wind was sweeping across the flat roof of the RCA Optics building at Cape Canaveral as nearly 90 newsmen milled about restlessly, most clad in old Army jackets and trenchcoats. "This is January 31, 1958," said a radio reporter into a tape-recorder microphone. "I am speaking to you from the United States missile test center."

Bathed in the blue-white glare of floodlights, the bulk of the Army's Jupiter C could be seen poised toward the heavens as if in supplication.

Edwin Tetlow, correspondent of the *London Daily Telegraph*, walked into one of the ten unpainted plywood phone booths, placed a call to London, and told his home office: "Hold the front page. It looks as if the bird might go up tonight!"

Soft-spoken, genial Kurt Debus, a member of the Peenemünde team who had fired more rockets than any man on earth, was in the blockhouse, calmly going through countless last-minute checks. Wernher von Braun, who had directed the development of Jupiter C, was waiting tensely in the telecommunications room of the Pentagon in Washington.

At 8:30 P.M., the countdown began. All systems were "go." At 10:48 P.M., Kurt Debus pushed the firing button and the Jupiter C, with a 30-pound satellite called Explorer I in its nose, filled the night with a thunderous roar as it slowly lifted from the pad.

"It's soaring beautifully!" a voice shouted in the blockhouse. "Go, baby, go!" called out one of the customarily blasé reporters on the roof of the RCA Optics building.

Listening to the verbal reports at the Cape, Wernher von Braun in the Pentagon was his normal cool and collected self. Inside, he felt a gut-wrenching sensation: the prestige of his adopted country was riding with the Jupiter C.

Soon the streaking missile was swallowed by the night. Now began a period of agonized waiting, until the big Goldstone radar in California picked up the man-made moon as it crossed the West Coast in its first orbit. Only then would it be known if the firing had been a success.

The seconds and the minutes ticked past. Ten minutes. Thirty minutes. Sixty. Ninety. Suddenly, an excited voice called out over the loudspeakers in the Cape Canaveral blockhouse and the Pentagon telecommunications room: "Goldstone has the bird!"

America's first satellite was in orbit.

Three minutes after the missile left its pad, NBC-TV commentator David Brinkley interrupted the premiere of *Comment*, a television panel show, to break the news. Twenty minutes later, the network presented a telephoned eyewitness report from its two men at Canaveral, Herb Kaplow and Roy Neal.

Then, while Kaplow remained behind to cover the Cape, Neal, clutching the precious film footage of the blast-off, leaped into an automobile and drove at breakneck speed ten miles to a private airfield, hopped into a chartered Piper Apache that was waiting for him, and flew to Jacksonville. After landing, he blew out a tire on the way into town, drove five miles on the wheel rim, and finally pulled up to NBC's affiliate, WFGA-TV. At 2:06 A.M., the first film of the launching was put on the NBC network, an hour ahead of rival CBS.

When the big news hit Huntsville, the lid blew off. It was 1:00 A.M. Sirens sounded in the streets to herald the lifting of the first U.S. satellite into orbit; cheering crowds jammed downtown sidewalks; police cars with horns and sirens at full blast roamed about aimlessly; fire engines sounded their alarms with no fire to go to. The place was bedlam.

In the courthouse square, hundreds applauded the burning of an effigy of Charles Wilson, the former secretary of defense who temporarily blocked the

Army missile program and was on the brink of closing down the Redstone Arsenal. Waving banners proclaimed: "Move Over, Sputnik" and "Our Missiles Never Miss."

Here, as friends and neighbors, lived the top scientists of Redstone—Wernher von Braun, Ernst Stühlinger, E. D. Geissler, among many others. Thousands of Huntsville people had worked on the program in one capacity or another. The Huntsville celebration succeeded beyond anyone's wildest dreams; it was estimated that at one time 10,000 people jammed the streets. "It's the greatest day in the history of Huntsville!" shouted a beaming Mayor R. B. Searcy above the tumult and the roar of the crowd.[4]

In his exile quarters in the Panama Canal Zone, Colonel John Nickerson, whom Huntsville credited for saving the Army missile program at the sacrifice of his own career, learned of the Jupiter C firing from a newspaper reporter. All Nickerson would say was that he was "under orders not to comment on missiles."

On vacation in Miami Beach, Charles Wilson was asked how he felt about being burned in effigy by the people of Huntsville. "I don't know why they're mad at me," he replied. "I'm the one who put them in the Jupiter business. They must have me mixed up with the Ku Klux Klan."[5]

Hardly had the American satellite made its first turn around the earth than a number of congressmen—all Republicans—popped up on local television programs around the nation, explaining the new scientific wonder, which few, if any of them, understood. There was a major flaw in the programs, however: they had been recorded in advance, on film, and they had the wrong satellite.

The five-minute videos had been made available to the lawmakers (at $5 for black and white, $14 for color) by the GOP Congressional Campaign Committee. Each legislator had dubbed in a movie clip of himself introducing and closing the piece before sending the film to his television station back home.

But the film had been based on the premise that the first American satellite would be fired by the Vanguard and would be spherical; instead, the Army's Jupiter C launched a satellite shaped like a bullet.

A little matter of touting on the tube the wrong missile, the wrong satellite, and the wrong distance from earth of the whirling object in space did not phase the Republican congressmen one iota. The next day, dozens of them besieged the TV studios of the House Office Building to get more film clips—discrepancies and all.

Wernher von Braun became an overnight American hero. His name, perhaps more than any other, had become synonymous with space. He was flooded by fan mail. In restaurants around Cape Canaveral and in Huntsville, excited waitresses, ignoring the other high-level missile-program officials seated at his table, crowded around von Braun to have him autograph menus.

Despite his Hollywood-type popularity, von Braun continued to dress simply and preferred to wear slacks and sports coats. He always looked to be overdue at the barber. His tanned, blue-eyed face gained him countless admirers, and his brisk step and weight-lifter's build belied his age of 45 years. Perhaps by coincidence, he made one concession to his superstar status: he traded in his weather-beaten old Chevrolet for a new Buick.

Although von Braun loved spontaneous conversations with those he met casually on the street, in stores, or at missile sites, he hated crowded cocktail parties. His longtime close friend, German rocket pioneer, Willi Ley, said of von Braun's disdain for cocktail parties: "Everybody wanted to get introduced to him so he didn't have a chance to get a drink."

Four days after Explorer I began beep-beeping around the earth, President Dwight Eisenhower was presiding at a routine meeting of Republican leaders in the White House. There was no specific topic, but the moon monopolized the conversations. Suddenly, the whole nation seemed to want an American to land on the lunar surface.

Only a week earlier, Nelson Rockefeller, who was running for governor of New York under the GOP banner, enthusiastically told Eisenhower that if the United States used nuclear explosions for propulsion, a vehicle could be launched that would fly to the moon and return. Rockefeller predicted that it would be "the most notable accomplishment of our time."

Later, Eisenhower scoffed at the proposal. "Some people think there are no limits to the amounts of money available for every conceivable project."[6]

Now the president's confidants, gathered around the White House conference table, also had moon dust in their eyes. When one suggested making a manned lunar flight a national goal, Eisenhower winced and replied, "I'd rather have a good [Jupiter C] than be able to hit the moon. We don't have any enemies on the moon!"[7]

Two weeks later, James R. Killian, president of the Massachusetts Institute of Technology, and Donald Quarles, the deputy secretary of defense, proposed an expenditure of $1.5 billion over the next few years to send a nuclear-powered rocket to the moon. Eisenhower brushed off the idea as "Buck Rogers fantasy."

Eisenhower, who was determined to cut governmental spending, regarded such "prestige schemes" as a waste of money and talent. Scientists around the nation were critical of the president's viewpoint. Eisenhower did not respond publicly, but to confidants in the White House he growled, "When a scientist gets before a television camera he gets excited. Hell's bells, I heard one of them talking about a shot to the moon—and we haven't even put up a full-sized satellite!"[8]

America was still basking in the glow of the 30-pound Explorer I when it was rudely jolted. On May 15, 1958, the Soviets orbited *Sputnik III*, a huge satellite weighing nearly one and a half tons. It was an awesome feat. U.S. intelligence reports from behind the Iron Curtain indicated that manned flights to the moon might be the ultimate goal of the Russian space program.

Americans were both dazzled and frightened by the Soviet space spectacular. A distraught Michigan telephone caller was soothed by an understanding desk officer at a Detroit police headquarters who assured her that the 3,000-pound *Sputnik III* would neither tilt the moon off course nor land in her backyard.

Congress and the Pentagon were deeply concerned over the military ramifications of the man-made moon, and a House committee summoned Wernher von Braun to testify on means for overcoming the Russians' huge lead in the space race. If the legislators expected him to solve the problem by waving a magic wand, they were doomed to disappointment.

"Our missile program must be backed by a large budget which permits its steady prosecution over a period of several years," von Braun asserted. "But even with no holds barred, I think it will be well over five years before we can catch up with the Soviets' big rockets because they are not likely to sit idly by in the meantime."[9]

As the grim congressmen pondered that jolting analysis, von Braun added, "There is a crying need for more money for basic and applied research . . . for development of bigger booster engines."

"Would you say that we are competing with the Soviets in space?" a senator asked.

"We are competing in spirit only," von Braun replied.[10]

Suddenly, space had become the big item in Washington.

Meanwhile, across the land, editorial writers, scientists, university intellectuals, Democrats, and Republicans were calling for the creation of a Department of Space, a proposal frowned upon by President Eisenhower. However, the satellite and moon-shot issue would not go away. At a White House meeting in mid-February 1958, Vice President Richard M. Nixon, a young Californian who had been hand-picked by Eisenhower to be his 1952 running mate, cautioned the president that the pressure to create a civilian space agency was enormous—and growing.

Eisenhower was dubious. He said that he had no intention of "getting involved in a pathetic race" and called a moon probe "useless."[11]

Six weeks later, as the drumfire continued, Eisenhower, a five-star general, realized that he had been defeated on the political battlefield, and he called on Congress to establish the National Aeronautics and Space Agency (NASA), which would have control of all space activities except those determined by a president to be primarily involving national defense.

On July 29, 1958, NASA was made official by law. T. Keith Glennan, president of Case Institute of Technology in Cleveland, was named the first administrator. Starting in name only, NASA grew rapidly by absorption. First, it took over the 8,000-man National Advisory Committee for Aeronautics (NACA), an agency that had been created prior to World War I for aviation research. Hugh L. Dryden, NACA's director, became NASA's deputy director.

John Hagen's 170-person Vanguard team and the 46-man Upper Atmosphere Sounding Rocket Group, under John W. Townsend, Jr., also were transferred to NASA. The Army-owned Jet Propulsion Laboratory at the California Institute of Technology, with 2,800 employees under William Pickering, became part of NASA's resources.

Rapidly NASA began gearing up for the daunting task of overtaking the Soviets' huge lead in space technology. Privately, large numbers of the American space experts were far from confident of achieving that goal.

# 20

# "I'm Tired of America Being Second!"

When ships to sail the void between the stars have been invented there will also be men who come forward to sail those ships.

—Johannes Kepler, 1610 A.D.

Seated at his desk in the Oval Office in late December 1958, President Eisenhower was irate. He had just learned that NASA Administrator T. Keith Glennan was about to publish civil service notices inviting astronaut applications. "Ridiculous!" Ike snorted between cigarette puffs. "Our military branches are filled with test pilots who can do the job."

It would prove to be a stroke of genius on Eisenhower's part. Ruled out were the matadors, scuba divers, mountain climbers, and publicity seekers. Instead, NASA would have a hard core of stable, experienced pilots who already had security clearances.

NASA painstakingly scrutinized the records of 508 test pilots and pared off those who failed to meet the minimum standards, including the height and age limitations (five feet, 11 inches and 40 years). Screeners were looking for men with such characteristics as intelligence without genius, knowledge without inflexibility, fear but not cowardice, courage without foolhardiness, self-confidence without egotism, physical fitness without being muscle bound, frankness without blabbermouthing, and fast reflexes without panic in a crisis.

NASA eventually narrowed the list of 508 test pilots to 32. These candidates were put through grueling physical tests, including telling how long a candidate could keep his feet in ice water. Psychological tests involved the most painful and often embarrassing self-scrutiny.

From the 32 survivors, America's first seven astronauts were named: Alan B.

Shepard, Jr., Virgil I. "Gus" Grissom, John M. Glenn, Jr., M. Scott Carpenter, Walter M. Schirra, Jr., Donald K. Slayton, Jr., and L. Gordon Cooper, all Navy, Marine, or Air Force officers.

These trailblazers would risk burning to death while perched atop a rocket waiting to be whisked into space, perishing in a midair explosion, succumbing to some yet unknown radiation, or suffering an agonizing death when an orbiting spacecraft could not be brought back to earth—for annual salaries not in excess of $12,770.

On December 18, 1958, a short time after President Eisenhower laid down the ground rules for selecting astronauts, Uncle Sam took a mighty leap to get into the space race with the Soviets when the Navy's 85-foot, 8,800-pound Atlas rocketed off the Cape Canaveral pad and circled the earth at a maximum altitude of 925 miles.

Two weeks after that feat, on January 2, 1959, the Russians shot a missile that zoomed past the moon, more than 240,000 miles away, and headed on into space where it became an artificial planet circling the sun.

Then, in April 1959, NASA unveiled a plan for Project Mercury, a program designed to harness the know-how of scientists, industry, and government to send an American into space. At the same time, NASA created a plan for a series of space-vehicle tracking stations to be located around the world. Each station would be equipped with large radio telescopes to pick up signals from orbiting American spacecraft.

Five leading corporations would be responsible for creating the integrated tracking network, a scientific innovation. Bell Telephone Laboratories was the lead company in the consortium, which also included Western Electric Company, International Business Machines, Bendix Aviation Corporation, and the New York–based engineering firm Burns and Roe.

"Our combined Mercury tracking team was headed by a Bell Telephone executive named 'Red' Goetchius," Kenneth A. Roe, chairman of the board of Burns and Roe, recalled years later. "He was an inspiring leader and an excellent diplomat. Red was chiefly responsible for our team's working together in harmony and successfully."[1]

Building the Mercury tracking network was a crash job. Engineers from the five firms involved hurried to farflung sites where stations would be constructed: Nigeria, Hawaii, Mexico, southern California, Bermuda, Zanzibar, Canary Islands, Australia (two stations), Cape Canaveral (primary control center), and aboard picket ships in the Indian and Atlantic oceans.

At each of these land sites, most of them isolated, on-the-spot studies were made of local conditions as they related to constructing the stations, siting electronic equipment, providing power utilities, and building access roads. This highly complicated tracking and ground-instrumentation system would help remove the term *man-in-space* from the pages of science fiction and put it in global newspaper headlines.

Because of the urgency, actual work on the tracking stations could not be staggered. Through Herculean effort, construction got under way simultaneously at all Mercury sites in June 1960. The network was believed to be the largest peacetime around-the-world project ever attempted at the same time. America's scientific and engineering know-how would be put to the acid test.

At the peak of activity, some 1,000 engineers and scientists, hundreds of subcontractors, vendors, and suppliers, along with perhaps 30,000 workers, were involved. Most of the workers were recruited at or near each site.

Mercury's tracking network had to be created with incredible precision. There was no margin for miscalculation. When a manned American vehicle eventually soared into space, its positions, speed, and direction would have to be monitored precisely. It would be necessary to establish and maintain communications with the astronaut in order to analyze his environment, condition, reactions, and spoken words.

''These critical requirements meant that the integrated tracking equipment would have to operate at high speed with maximum reliability and uncanny accuracy, around the clock and in any weather conditions,'' Burns and Roe's Kenneth Roe recalled. ''Each radio antenna at each site had to be critically established between other antennae throughout the global network.''[2]

Logistics, in the words of Roe, were ''mind boggling.'' Such critical equipment as radar, generators, switchgear, and antenna towers, shipped from all parts of the world, had to be at the designated site at precisely the right time. Coordination was crucial. Each site had to be ready for experts of the five-company consortium to rush in and install their specific pieces of equipment. There could be no lost time, no overlapping of functions.

Construction headaches grew more severe when especially remote regions and routes were involved. At one primitive locale in North Africa, armed pickets had to be put out to guard against attacks by wild beasts.

The massive job was so complicated that 34 different operating and maintenance manuals had to be created, a tedious and demanding task in its own right, for there were no precedents.

In an astonishing display of American technology and engineering know-how and teamwork, the tracking stations were ready for acceptance tests within 18 weeks.

Meanwhile, Uncle Sam's first seven astronauts had soon discovered that training was the name of the game. They trained until it seemed that the rigorous labors of Hercules were mere child's play—how to kill and eat a snake in the jungles of Panama; how to walk through volcanic lava in Hawaii; how to make a tent out of a parachute in case one came down in the African desert.

Perhaps to the chagrin of the astronauts, the first Project Mercury ''Americans'' to leave the earth's atmosphere were a seven-pound monkey named Able and a one-pound squirrel monkey named Baker—''monkeynauts,'' they were tabbed by the press. Both heavily wired to gather data about the effects of space flight on living beings, Able and Baker were plucked alive, hale and hearty, from the Atlantic Ocean on May 28, 1959, after a 1,700-mile trip that took them to a height of 360 miles in the nose of a Jupiter C missile.

By 1960, NASA had created the framework for a mission to the moon (Project Apollo), and the agency asked President Eisenhower for approval to shoot for a 1969 target for lunar orbit. Eisenhower's Scientific Advisory Committee studied NASA's proposal and concluded that a moon landing would be prohibitively expensive. On that ground, the president turned thumbs down on Apollo, but the United States would continue with the Project Mercury space exploration.

Although NASA was taking great strides in creating a unified assault on space, there was still a major need for scientists and engineers with long experience in the field of missiles. So the Wernher von Braun/General John Medaris team at the Redstone Arsenal was transferred to NASA on July 1, 1960, and the installation became the George C. Marshall Space Flight Center with von Braun as its director.[3]

Von Braun's mission was to develop an efficient and reliable system for lifting multiton loads into orbit around the earth and into deep space. Toward this end, he and other members of his team at Huntsville would develop the powerful Saturn rocket.

A few members of the Peenemünde team had left government service to accept far more lucrative jobs in private industry. No Peenemünde expert (including von Braun) was making more than $10,000 annually as civil service employees. Others on the Overcast group had returned to Germany. Yet 89 Peenemünde men remained.

In the months ahead, von Braun, with his customary driving energy, traveled almost constantly between Huntsville, Cape Canaveral, and Washington. In so doing, he had become almost a stranger to his two young daughters and son, Peter, one of whom told him she wished he worked in a drugstore so he would be home at night.

A pleasant fallout from his visits to the Cape was that he found time after his 14-hour workday and on weekends to participate in his favorite sports in the beautiful, clear, and warm waters off the coast of Florida. He was an avid skin diver and loved to swim, motorboat, and fish. Once he told reporters that he was going to retire soon and do nothing but go skin diving.

As a celebrity, von Braun was besieged by flocks of reporters and television camera crews. Once, walking away from an outdoor press conference at the Cape, while the orange cloud of smoke from a blast-off still hugged the ground, von Braun smiled broadly at his latest success, brushed back an errant shock of graying blond hair, and recalled something Jules Verne had said: "Anything one man can imagine," said Verne, the Frenchman who more than a century earlier wrote of men flying to the moon, "other men can make real."[4]

It was typical von Braun, an articulate and often eloquent man. He liked to compare space exploration to Columbus, the Wright brothers, and Charles Lindbergh. When he made an analogy he liked especially well, he would break into a slightly lopsided smile, arch his eyebrows, and look to make certain that the listener had gotten the point.

Some 25,000 people, bundled in heavy coats, gloves, and woolen scarves, massed before the east portico of the Capitol in Washington, D.C., on January 20, 1961, to witness the inauguration of American history's youngest president, John F. Kennedy. Among those present on the platform were former President Harry Truman, outgoing President Dwight Eisenhower, newly elected Vice President Lyndon B. Johnson, Jacqueline Kennedy, and Lady Bird Johnson, as well as singer Marian Anderson and poet Robert Frost.

At 12:30 P.M., Jack Kennedy, bareheaded and wearing no topcoat despite the near zero windchill, delivered his enduring "ask not what your country can do for you" address.

Since the days of George Washington, presidents have been hard put to find something to keep a vice president happy and gainfully occupied, for the number-two man's primary function is to stand by in the event that the president is killed or dies. Johnson, a hard-driving, hard-cussing politician from Texas, did not wait for Kennedy to dig out a job for him. Johnson had just the post in mind—overseer of NASA.

Lyndon Johnson, who had come to Washington as a congressman in 1937 with minimal formal education and owning little more than the clothes he was wearing, knew little about space and aeronautics, but riding herd on NASA would give him a portion of the enormous clout he had as majority leader of the Senate prior to becoming Jack Kennedy's running mate. Johnson had been the most reluctant of any of America's vice presidential candidates. If he and Kennedy triumphed, Johnson would have to give up his enormous power. Worse, he would fall almost totally out of the limelight he so loved. Johnson was fond of quoting the old cliché: "A mother had two sons. One went to sea, and the other became vice president of the United States, and neither was ever heard from again."

Jack Kennedy had surrounded himself with an elite clique, mostly from the hallowed halls of Harvard and Yale, and its members were aghast at the unthinkable specter of the rough-hewn cowboy from Texas gaining too much authority in the administration. So the president turned thumbs down on Lyndon Johnson's pitch for power as overseer of NASA but delivered up the chairmanship of something called the Space Council—the perfect spot, Kennedy felt, for his veep.

When Congress passed the Space Act in 1958, it designated the president as chairman of the Space Council. But Ike Eisenhower had shunned the post, in the words of an aide, "like the Devil shuns Holy Water." Jack Kennedy was interested, but it better served his purposes to put Lyndon Johnson in the job. After all, a vice president had to do something.

No one seemed to be quite certain what the Space Council was supposed to do. So Johnson, no bashful bride, promptly set down the duties of the new Space Council chairman—himself: "He is to advise the president of what this nation's space policy ought to be."

Johnson wasted no time in presenting his first bit of advice to his far younger boss, Jack Kennedy. While in the Senate, Johnson said, his committee had urgently advocated improving America's space propulsion capability with what he called "the big booster," meaning a rocket of the type Wernher von Braun and his Peenemünde team at Huntsville were hoping to develop. Ike Eisenhower had vetoed the project, Johnson said, but he urged Kennedy to give the big booster a trial.

Almost at once, Johnson called Space Council hearings to review the U.S. space program as a whole. The mass hysteria of Sputnik had largely dissipated, but Uncle Sam had yet to produce anything nearly as dramatic. Kennedy apparently had been impressed by part of the testimony presented at the hearings, for on April 20, 1961, he wrote Johnson a letter, asking him what project the United States could undertake that had a good chance of beating the Russians' space spectaculars.

Lyndon Johnson, a man of enormous energies, returned to the Oval Office on May 5, carrying a large manila envelope containing the first major report of the Space Council. No doubt it reflected the vice president's own views. The best possibility for the United States, the report emphasized, was a project to land a man on the moon.

While the Russians had the propulsion capability for putting a multimanned spacecraft into orbit, the Space Council declared, they were probably not much closer than the United States to actually landing a man on the moon and bringing him back. If Kennedy's goal was to "win" within the foreseeable future, then the lunar landing was the likeliest project, the report concluded. President Kennedy, Johnson felt, was pleased and excited about the Space Council recommendations, but he was noncommittal.

Once the Space Council report was submitted, Lyndon Johnson seemed to lose interest in the chairmanship. What else was there for the panel to do? Since it had little else to accomplish, he could not even lash it on to faster movement—driving a group forward being his specialty.

A few months earlier, shortly after being sworn into office, President Kennedy sent for 54-year-old James E. Webb, a North Carolinian who had been budget director and undersecretary of state in the Truman administration. Now he was a director and assistant to the president of Kerry-McGee Oil Industries in Oklahoma and a director of McDonnell Aircraft Corporation in St. Louis. Much to Webb's astonishment, Kennedy offered him the post of administrator of NASA.

Webb, who had served on active duty as commander of the 1st Marine Air Warning Group in World War II and had held key executive posts at Sperry Gyroscope Corporation, told Kennedy that he was without expertise in space science and technology and advised the president to tap an engineer or scientist for the job.

Kennedy refused to take "no" for an answer, pointing out that rockets were becoming so powerful that they could open up "the new ocean of space" and that the skipper to sail those waters should be someone with top-level administrative experience like Webb. Vice President Lyndon Johnson, renowned as an "arm twister" when he had served as Democratic floor leader in the Senate, got into the act. Webb's experience with high-technology companies in the business world would be invaluable as NASA chief, Johnson declared.

The powerful combination of Kennedy-Johnson was too much to overcome. Webb accepted the post. The native of Tally Ho, North Carolina, landed on NASA headquarters in Washington with both feet running. Webb and his key aides huddled to determine the adequacy of NASA's long-range plans and found that sufficient priority and funds were woefully lacking.

So Administrator Webb called on President Kennedy in the White House and urged him to reverse an earlier decision to postpone Project Gemini, the manned space flights that were planned as a follow-up to the Mercury program. Earlier, Kennedy had approved NASA funds for larger rocket engines but not for development of a new generation of man-rated boosters and manned spacecraft.

The gist of Webb's plea was: what good are powerful rockets if they don't take man into the heavens to explore the universe? "When President Eisenhower, in the 1962 budget, reduced NASA's request by $240 million and specifically eliminated funds to proceed with manned space flight beyond Mercury, it emasculated our ten-year plan before it was even one year old," Webb explained. "Unless that decision is changed, it guarantees that the Russians will, for the next five to ten years, beat us to every spectacular exploratory flight."[5]

Seated in the rocking chair kept in the Oval Office because of his chronic ailing back, the president puffed on a cigar, listened intently, and made no reply. Feeling that he had made an impression on Kennedy, Webb continued: "Mr. President, the first priority of this country's space effort should be to improve as rapidly as possible our capability for boosting large spacecraft into orbit with rockets, since this is our greatest deficiency."[6]

Jim Webb soon learned that Kennedy had not bought his sales pitch. The president's decision was to proceed cautiously in funding the space program. Webb was disappointed, telling confidants that Kennedy was not moving toward a resolution of the great international policy issues with regard to space exploration about which he had spoken so eloquently when he was recruiting the new NASA boss.

Meanwhile, a limited space program proceeded. On January 31, 1961, 11 days after John Kennedy took over the White House, a souped-up Redstone rocket launched a Mercury capsule weighing 2,400 pounds to an altitude of 155 miles and a distance of 420 miles. It was precisely three years since Wernher von Braun and his team put America's first satellite into orbit.

On board the Redstone for the 18-minute ride was Uncle Sam's first astrochimp—a 37-pound chimpanzee named Ham—who was fed tablets that tasted like bananas during his breathtaking flight. Ham returned in good shape and found himself to be a hero of sorts. True to his name, the chimp, finding himself the center of attraction, mugged shamelessly for a raft of television and still cameras.[7]

Four thousand miles from Cape Canaveral on the morning of April 12, 1961, an excited announcer broke into Radio Moscow programming with a startling revelation: the Soviet Union had put the first man into space. Tass, the official Soviet news agency, reported that 27-year-old Air Force Major Yuri A. Gagarin, a smiling, boyish-faced man, had circled the earth in a 10,395-pound *Vostok I*. The spacecraft's orbit had a maximum altitude of 187.75 miles and each revolution around the earth took 89.1 minutes.

Gagarin applied a braking device and landed in "the prescribed area" of the Soviet Union. Tass reported that the first man in space had said, "I feel well and have no injuries or bruises."

If Yuri Gagarin felt well, then NASA and its 9,000 scientists, engineers, and technicians were feeling ill. Russia had put a man into orbit, and the best that the United States had been able to achieve was to send a chimpanzee and a pair of monkeys into suborbital flight. President John Kennedy voiced the frustration felt by tens of millions of his countrymen: "I'm tired of America being second in space!"

The Soviet Union wasted no time in cashing in on the propaganda impact of the Gagarin show-stopper. In his first telephone conversation with the cosmonaut, Premier Nikita Khrushchev, a bombastic type, boasted, "Let the capitalist countries catch up with our country!"

A day later, *Pravda*, the Communist newspaper in Moscow, trumpeted: "In this achievement are embodied the genius of the Soviet people and the powerful force of communism. The Gagarin flight was evidence of the virtues of victorious socialism." *Pravda* added that the space feat was "evidence of the global superiority of the Soviet Union in all aspects of science and technology."

In the White House, President Kennedy was seething over the Soviet boasts. Pacing the Oval Office, he told confidants that the U.S. space program would be used as an instrument of national strategy, rather than primarily as a means for scientific research of the universe. Translation: Uncle Sam and Ivan the Bear were locked in a race for space.

# 21

# An Astronaut Is Missing

Three weeks after Yuri Gagarin's historic flight, the United States was ready to surge back into the space sweepstakes. Thirty-seven-year-old Commander Alan Shepard, described by his former coach at the Naval Academy as "a real hard-nosed kid with a lot of guts," would be the first American to soar into space. Target date for the blast-off was May 5, 1961. When selected as one of the original seven astronauts, Shepard's reaction was: "My feelings are quite simple. Without being too Navy Blue and Gold, I'm here because it's a chance to serve my country. I'm here, too, because it's a great personal challenge."

Shepard dismissed the dangers of space exploration with a shrug. "You've got a better chance of coming back safely from outer space than you do in trying to cross Pennsylvania Avenue in Washington during the rush hour," he explained with a tad of exaggeration.

On the day of the Big Flight, Shepard, a wiry 165-pounder, was awakened at 1:05 A.M., ate a low-residue breakfast of steak and eggs, and underwent a final physical checkup by a battery of medical specialists. Instruments were attached to his body to measure respiration, heart beat, and body temperature. He then climbed into his 32-pound nylon spacesuit, joking good-naturedly with his fellow astronauts. They quipped that if Shepard wanted to back out, they would be happy to take his place.

A round of handshakes, and Shepard left the checkout building at 5:22 A.M. and calmly entered the ten-foot, 2,800-pound Mercury capsule named *Freedom 7*, perched atop a Redstone missile at the launching pad. NASA, an unromantic type, called the trailblazing event Test Number 108.

Finally, at 9:34 A.M., a burst of flame launched the rocket off its pad and the flight was underway. Tens of millions of Americans watched the spectacle on television. At T plus 142 (two minutes and 22 seconds after lift-off), 37 miles above the earth, the space capsule separated from the Redstone and was on its own.

Approaching the maximum altitude of 115 miles at T plus 270, Shepard looked through the periscope and observed a panorama 1,600 miles wide, extending to the Carolinas and the Bahamas. No American had previously seen that sight. Millions of awed TV watchers heard him exclaim: "What a beautiful view!"

A few minutes later, the astronaut called out to Mission Control at the Cape: "Everything A-Okay!" It was television drama at its best.

Everything also functioned A-Okay in the global tracking network. *Freedom 7* was monitored precisely, from lift-off to capsule splashdown in the Atlantic near the Bahamas, 302 miles from Cape Canaveral. Although America's first man in space had traveled only a fraction of the distance logged by the Soviet astronaut in *Vostok I*, and his capsule had a top speed of only a quarter as great as Gagarin's, Shepard had maneuvered his craft in space by firing small rockets, an achievement the Russians could not claim.

Americans were jubilant. National pride had been restored. Alan Shepard, Jr., son of a retired Army colonel, was given the hero's treatment. A cheering throng of 250,000 persons lined Pennsylvania Avenue in Washington to welcome the astronaut.[1]

On Capitol Hill, Senator J. William Fulbright of Arkansas hailed the flight as "the most encouraging accomplishment in technological affairs" and as a setback for "America's defeatists."

As a result of the enormous publicity over Shepard's space spectacular, Wernher von Braun, whose Redstone team had developed the rocket, was flooded with speaking requests. Suddenly, America was mesmerized by space. Von Braun had time to accept only one in a hundred of the invitations, but his speeches resulted in what media people call "good copy," and his humor delighted audiences.

In late May 1961, before his speech began at a Wisconsin Manufacturers Association dinner in Milwaukee, the banquet hall electrician made sure the microphone was in working order: "Testing—one, two, three, four." When von Braun got up to speak, he pulled a space age switch that drew the night's biggest laugh, chanting into the mike: "Testing—five, four, three, two, one."

Five days later, von Braun was in Decatur, Alabama, where he was presented with a bronze plaque naming him "1961 International Boss of the Year." The honor came from the National Secretaries Association on the recommendation of his own secretary, Bonnie Holmes. Much to the missile pioneer's embarrassment, his exemplary qualities, as cited by Holmes, was read to the audience: "He very rarely scolds and never uses profanity. He always takes time to be nice to people working under him. He doesn't dictate too fast and his English is so good that I don't ever have to correct it. And he always knows exactly what he's going to say."[2]

In May 1961, an American navigational satellite inadvertently crashed in Communist Cuba, causing no damage but killing a cow. Fidel Castro, the bearded Cuban revolutionary who had been feuding with the United States, sought to milk propaganda mileage from the incident. To dramatize "Yankee imperialistic aggression," he ordered that a state funeral be conducted for the departed cow, which, in death, had become a Cuban national hero of sorts.

When the national jubilation and buoyancy subsided after President Kennedy's challenge before Congress to "fly man to the moon in this decade," NASA officials began taking a hard, sobering look at the task to be confronted. Could it really be done? Even the concepts of manned space flight were only three years old, and traveling in space over vast distances was still merely a dream. Giant new rockets were needed, and their ability to launch on time had to be perfected, since going to the moon required the accurate hitting of "launch windows."

Man himself was a great unknown. How would astronauts be affected by prolonged weightlessness, possible solar radiation, and suspected meteroid hazards? Would a man in space lose his mental faculties due to incredible psychological stress? Could a human being really function on a two-week mission that would require finite concentration and intricate maneuvers? NASA had only Alan Shepard's 15-minute flight on which to base its knowledge. All these questions involved problems of unknown dimensions.

Although astronomers and scientists had studied the moon for centuries, that planet, too, remained a great unknown. Whether the lunar surface would be firm enough to support a landing craft was open to question.

Finally, a master plan had to be created for a gargantuan task that would rival in scope the Manhattan Project that developed the A-bomb in World War II. A skilled and dedicated government space team had to be built that would work closely with industry and universities in the development of rockets, spacecraft, and massive facilities for testing and operations. All of these formidable problems had to be overcome, with few precedents to light the way—and they would have to be solved in the relatively short period of nine years.

Even while NASA's leaders were wrestling with the awesome problems confronting them, the United States and Russia were continuing to pierce the upper atmosphere with moon-shots. Scientific theories were rampant in both camps about what the first manned spacecraft to reach the moon would encounter.

Back in 1955, T. Gold, a noted British astronomer, published a paper in which he claimed that the lunar surface was dust and that any vehicle that tried to land there would sink out of sight, taking its crew to death by agonizing suffocation. This startling conjecture had been taken quite seriously by both American and Soviet space planners.

Consequently, beginning on August 17, 1958, the United States launched a series of rockets aimed at hitting the moon, an effort that continued until mid–1961. All either blew up, fizzled, or missed the target by 37,000 miles. During this time frame, the Soviets also were aiming rockets at the moon, and Lunik 2 finally scored a bull's eye—the first man-made object to crash on the lunar surface.

In June 1961, Captain Virgil "Gus" Grissom, a 35-year-old Air Force test pilot and Korean War combat ace, was selected to be the second American to make a suborbital flight. A native of Mitchell, Indiana, Grissom's decision to become an astronaut had not been reached without certain misgivings.

"Occasionally, I lie in bed at night and think, now what in the hell do I want to get up in that thing for?" he told a reporter. His own answer was a patriotic one: "If my country has decided that I'm one of the better qualified people for the mission, then I'm glad I can participate."

On July 31, after a 16-minute flight that carried Grissom 116 miles into space, his capsule splashed down in the Atlantic 303 miles from Cape Canaveral. Before he could climb out, explosive bolts accidentally blew out the side hatch of the capsule. The capsule flooded, and Grissom, clad in his bulky spacesuit, had to swim for more than two minutes before a hovering helicopter plucked him from the sea. Grissom was "a bit uneasy" but suffered no harm.

Now NASA was ready to take a major leap forward and selected Marine Lieutenant Colonel John H. Glenn for a mission to orbit the earth. At age 40, he was the oldest of the original astronauts.

A native of Cambridge, Ohio, where he was an honor student in high school and won letters in football, basketball, and track, Glenn played the trumpet, sang in the church choir, washed cars for pocket money, and worked as a lifeguard at a summer camp.

Glenn was deeply religious since boyhood. He had belonged to a group called the Ohio Rangers, whose youthful members vowed never to use profanity. One evening Glenn and a close friend, Edwin Houk (much later a minister), were with a group that began singing boisterously, *Hail, Hail, the Gang's All Here*. Houk continued with the phrase, "What the hell do we care."[3] Young John Glenn became angry over the flagrant transgression committed by his fellow Ohio Ranger and came as near to cussing out his pal Houk as one could come without actually cussing.

Glenn won his wings as a Marine aviator in March 1943 and flew 59 combat missions during the bloody fighting in the Marshall Islands. A decade later, he flew 90 missions in the Korean War.

Once chosen to be an astronaut, Colonel Glenn drove himself like a man possessed to prepare for the Big Day. He ran two miles every morning and dieted to bring his weight down from 195 to 168 pounds. His evenings were spent in the study of books, charts, and maps.

Glenn's flight was postponed for ten days because of bad weather and poor conditions in the recovery-site locale. Finally, at 9:47 A.M. on February 20, 1962, his spacecraft *Friendship 7*, was blasted into orbit by an Atlas-D rocket.

Nine hundred miles north of the Cape, President Kennedy and Vice President Johnson were in the Oval Office of the White House, their eyes glued to the television screen, watching *Friendship 7* soar into space. Johnson, critics charged, never went to the restroom without considering the political ramifications of that ritual. So now, aware that the Kennedy-Johnson victory had resulted partly from the ticket's reaching out to minority groups, the vice president suddenly turned to Kennedy and said, "If John Glenn were only a Negro!"

Through the miracle of the sophisticated electronic equipment at the tracking stations along the path of *Friendship 7*, millions of television viewers heard Glenn describe his night view of Perth, Australia, whose residents had turned on their house and building lights for the historic event.

Experiencing some difficulty with the spacecraft's altitude control system, Glenn switched to manual controls and reported it to be "smooth and easy." After considering whether the flight should be cut short, Mission Control in Houston gave a go-ahead for a full three orbits as planned. Descending safely, *Friendship 7*, having been tracked every foot of its flight, landed close to the destroyer *Noa*, the recovery ship in the Atlantic, at 2:43 P.M.

Glenn's three-orbit flight had lasted four hours and 56 minutes. Those who pulled him from the capsule as it bobbed in the ocean anticipated that the first American to orbit the earth would speak an immortal phrase that would forever ring in the annals of history. Instead, Glenn's first words were: "Boy, it was hot in there!"[4]

America hailed John Glenn much as it had Charles Lindbergh, "the Lone Eagle," nearly 35 years earlier. A modest, yet self-possessed man, Glenn was showered with honors. Shortly after news of his safe landing became known, the U.S. Post Office Department issued a blue and yellow four-cent stamp captioned "U.S. Man in Space."

Plans were rapidly made for parades in New York City, Washington, D.C., and Glenn's hometown, New Concord, Ohio. When Lyndon Johnson learned of the impending ticker tape parade down Broadway in the Big Apple, an extravaganza that would draw the focus of the world and its media, he asked permission to accompany John Glenn.

President Kennedy and his tightly knit White House clique were dubious: why should a vice president be injected into the act? Undaunted, Johnson responded that he would not be participating as vice president and convinced Kennedy that it was right and proper for him, as chairman of the Space Council, to ride prominently in the New York City affair.

John Glenn, as with most of the original seven astronauts, was earning a salary in the $10,000 range; his five-hour trek through the perils of space entitled him to extra flight pay of $245 for that month. Each astronaut had a wife and, in most cases, children to support. So a few months earlier, the seven men sold *Life* magazine exclusive rights to their first-person stories and those of their families for $500,000—quite a tidy sum at the time. That arrangement, while approved by NASA and perfectly legitimate, brought down on their heads the wrath of large portions of the American press, which objected to being excluded.

Part of the half-million dollars was collectively invested by the seven astronauts in a Florida luxury motel at Cocoa Beach; a middle-income apartment project in Washington, D.C.; and a motel on Grand Bahama Island, in the Caribbean.

As a result of the barrage of criticism, the Cocoa Beach property, the Cape Colony Inn, was later offered for sale.

On the heels of John Glenn's feat, preparations promptly got under way for launching the second Mercury mission. Lieutenant Colonel Donald K. Slayton, a steel-nerved pilot who got his wings at age 19 and flew 56 bomber missions in the Mediterranean theater before he was 21 and then seven combat sorties in the Pacific, was the logical choice to be the second American to orbit the earth.

Then Dame Fate frowned on Deke Slayton, a dairy farmer's son from Sparta, Wisconsin. Doctors discovered that he had a slightly irregular heartbeat, and he was grounded. Slayton's six colleagues appealed directly to President Kennedy to overrule the group of doctors. Kennedy knew a hot potato when one was thrust into his hands, so he handed it off to Vice President Lyndon Johnson, who invited the seven astronauts to the LBJ ranch in Texas to thrash out the matter. Johnson, always the consummate politician, listened intently to the pleas of the six astronauts to reinstate the popular Deke Slayton.

Everyone had a wonderful weekend at the sprawling spread in Texas. Nothing was ever done about Slayton's grounding. So his six colleagues ''elected'' Deke their leader, thereby conferring upon him immense power in future NASA decision making.

In the fall of 1961, the thousands of pieces in the Apollo master plan were beginning to fit together. However, NASA leaders were confronted by perhaps the most crucial decision to be reached in the moon-flight program. Like wagonmasters of America's Old West steering their horse-drawn caravans into largely unexplored territory, NASA had to choose the best path to the moon and back. It was an explosive issue, with skilled and dedicated scientists, in and out of NASA, stridently voicing divergent opinions.

Three courses were considered. The first course was the direct approach, which would have involved the fewest steps and the fewest parts and supplies, but it would have required a rocket far more powerful than any yet designed to boost the spacecraft from earth directly onto the moon. This technique had ardent and powerful supporters, including Jerome Wiesner, President Kennedy's scientific advisor, and Brainerd Holmes, who Jim Webb had chosen to head Apollo in Washington.

The second course was the earth-orbital-rendezvous (EOR) mode in which the launch problem would be solved by using two rockets, one to propel the spacecraft and another, launched in a slightly lower orbit, to carry extra fuel. When both were in orbit, the velocity of the fuel-carrying rocket would be increased to bring its orbit up to that of the higher vehicle, a maneuver called a Hohmann transfer. When the two orbiting rockets were side by side, the extra fuel would be jettisoned and the vehicle carrying the astronauts would make a landing directly onto the lunar surface.

The third course, a lunar-orbit-rendezvous (LOR) mode was championed by

John Houbolt, chairman of a study group at NASA's Langley Research Center in Virginia. This procedure would be less expensive and more reliable by employing a separate moon-landing vehicle instead of taking the entire spacecraft onto the lunar surface, as would be the case with the other two schemes, Houbolt argued. Houbolt reasoned that the main spacecraft could be set in lunar orbit. One man would remain in the orbiting vehicle while two astronauts descended to the moon in a ferry that would later ascend with them to rejoin the mother ship in an intricate rendezvous operation.

When Robert E. Gilruth, director of the new Manned Spacecraft Center in Houston, heard of the Houbolt plan, he was convinced that this was the way to go. The LOR technique also had the strong backing of Wernher von Braun at the rocket center in Huntsville. In December 1961, Gilruth and his top associates flew to Washington and made an earnest appeal to Brainerd Holmes to approve the lunar-orbit-rendezvous mode, but Holmes remained unconvinced.

A behind-the-scenes hassle simmered for months, and finally the LOR was approved by NASA's Jim Webb, who shuttled the plan on to Jerome Wiesner and his panel of presidential science advisors. Wiesner and a few of the others dug in their heels in opposition, and the hot potato was sent up the chain of command to the Oval Office in the White House. Six months after the squabble had erupted, John Kennedy approved the LOR technique for landing on the moon.

Even before the high-level moon-landing dispute was settled, Lieutenant Commander Malcolm Scott Carpenter, a native of Boulder, Colorado, was preparing to follow John Glenn into earth orbit. His mission would drive home the fact that astronauts were engaged in a profession rife with hazards and the threat of sudden death.

Carpenter always had been frank about his reason for volunteering to become an astronaut—"a chance for immortality." Most men never have that chance, he declared. "This is something I would willingly give my life for," Carpenter told the *New York Herald Tribune*.[5]

On May 24, 1962, Carpenter completed a three-orbit flight in the space capsule *Aurora 7* and reentered the atmosphere at 12:32 P.M., overshooting the impact area. Nine minutes later, *Aurora 7* splashed into the Atlantic Ocean 138 miles northeast of Puerto Rico—some 250 miles beyond the point where it was to have landed.

For 45 minutes, near-panic gripped Cape Canaveral: Carpenter and his capsule could not be contacted by radio and apparently were lost. If he had landed in the water, his locale was unknown. Millions of television viewers all over the world waited nervously, hoping for the best, expecting the worst. Scott Carpenter was not too enthused about his predicament either. Search planes, ships, and helicopters scoured the region for hundreds of miles in each direction before a Navy aircraft spotted the astronaut, haggard, soaked to the skin, but calm and collected, bobbing in a life raft beside the floating capsule.

As soon as Commander Carpenter was aboard the USS *Intrepid*, a telephone call from President Kennedy was waiting. "You did a great job," Kennedy enthused. Replied Carpenter: "Mr. President, I'm sorry my aim wasn't a little better."[6]

As Wernher von Braun had predicted in his testimony before a congressional committee in the wake of Sputnik, the Soviets continued to gain global headlines for their space spectaculars. On August 11, 1962, Major Andrian Nikolayev was launched in *Vostok III*, and began to orbit the earth; a day later, Colonel Pavel Popovich blasted off in *Vostok IV*, entering a similar orbit.

For the first time, two spacecraft were going around the earth together, at one point no more than three miles apart. Four days later, the two capsules landed safely, about 150 miles apart and within six minutes of one another. Both had traveled over one million miles.

Less than a month after the Russians' twin-orbit caper, President John Kennedy and a sizable entourage lifted off in the presidential jet from Andrews Air Force Base, outside of Washington, D.C., on the first leg of a two-day tour of four space installations. Aides at the White House billed the president's jaunt as a "business trip" preliminary to the drafting of the budget for the next fiscal year and insisted that not the faintest aura of politics was involved.

Capitol insiders had a different view: Kennedy was taking the headline-grabbing trek to shore up his political base and to justify the expenditure of tens of billions of dollars to put an American on the moon.

# 22

# "But Why, Some Say, the Moon?"

President Kennedy arrived at the George C. Marshall Space Flight Center at Huntsville on the morning of September 11, 1962. Among those on hand to greet him were Wernher von Braun and Major General Francis J. McMorrow, head of the Army Missile Command. Dapper in a dark blue, pin-striped suit, the president carried a hat in one hand. Kennedy had always hated hats, but someone once said that a president should be seen with a hat, so he bowed to the tradition—halfway.

Those in Kennedy's traveling troupe included NASA's James Webb and Vice President Lyndon Johnson. Never overburdened by humility, Johnson was uncomfortable playing second banana to the youthful, charismatic president. Hardly had Kennedy stepped from the presidential jet at Huntsville than he was besieged by a jostling flock of reporters and cameramen, while Johnson stood alone on the fringes.

In the huge assembly building where von Braun and his team were developing a powerful Saturn, President Kennedy was studying the rocket intently when loud voices erupted nearby. Clearly nettled, he pretended not to hear as Jerome Wiesner, his principal scientific advisor, and von Braun engaged in an argument over the best technique for putting a man on the moon.

When Kennedy finished inspecting the Saturn, von Braun turned to the president and said: "This will fulfill your promise to put a man on the moon by the end of this decade—and by God, we'll do it!"[1]

Accompanied by von Braun, President Kennedy and his party flew to the Launch Operations Center at Cape Canaveral, where a rocket-firing demonstration was laid on for the visiting dignitaries. Then the Kennedy entourage climbed back into the presidential jet and winged toward Houston.

Since the beginning of Mercury, space flights were directed from a control center at Cape Canaveral. However, Congressman Albert Thomas, an influential Democrat who represented the Houston district, used his heavy clout on Capitol

Hill to get funds appropriated for constructing in Houston a mammoth facility to be known as the Manned Spacecraft Center. Flight direction would be moved from the Cape to Houston.

President Kennedy arrived in Houston after dark and promptly got in some political plugs before a bevy of newsmen and camera crews. Flanked by Congressman Thomas, who came to Washington in 1937, and Lyndon Johnson, the president reminded Houstonians that the Manned Spacecraft Center "will become the heart of a large scientific and engineering community."

Kennedy pointed out that during the next five years, NASA expected to double the number of scientists and engineers in the Houston area, to increase its outlays for salaries and wages in the region to $60 million a year, to invest some $200 million in plant and laboratory facilities, and to contract for new space efforts over $1 billion from the new center in Houston.

Lyndon Johnson told reporters: "I'm going to see that Houston gets its fair share of the space budget."[2]

The *Houston Chronicle* carried a double, eight-column banner headline: "PRESIDENT PREDICTS SPACE CENTER TO BOOM INDUSTRIAL SOUTHWEST." It was the kind of media coverage that would bring joy to the heart of any politician. John Kennedy was no exception.

On the second day of his tour, the president was greeted with thunderous cheers by a throng of 40,000 when he arrived under a broiling sun to give a speech at Rice University Stadium in Houston. Most of those in the audience were students.

"No nation which expects to be the leader of other nations can afford to lag in space exploration," Kennedy declared. "Those who came before us made certain that this country rode the waves of the modern inventions . . . and this generation does not intend to founder in the backwash of the coming age of space."[3]

Rice students could relate to the upbeat note—and to the young president. They cheered raucously.

Kennedy, with perspiration streaming down his handsome face, paused and then departed from his prepared text: "We mean to be part of it—we mean to lead it!"[4] A roar from 40,000 throats echoed across Houston.

Kennedy continued with one of the longest sentences on record:

> If I were to say that we shall send to the moon, 240,000 miles away from the control station in Houston, a giant rocket more than three hundred feet tall, the length of this football field, made of new metal alloys, some of which have not yet been invented, capable of standing heat and stresses several times more than have ever been experienced, fitted together with a precision better than the finest watch, carrying all the equipment needed for propulsion, guidance, control, communications, food and survival, or an untried mission, to an unknown celestial body, and then return it safely to earth, reentering the atmosphere at speeds of over 25,000 miles per hour, causing heat about half that of the temperature of the sun, almost as hot as it is here today, and do all this, and do it right, and do it first before this decade is out, then we must be bold.[5]

Then the president asked rhetorically: "But why, some say, the moon? Why choose this as our goal? And they may well ask, why climb the highest mountain? Why thirty-five years ago fly the Atlantic?"

Again Kennedy departed from his text: "Why does Rice play Texas [a traditional foe]?" The students erupted with shouts and applause.[6]

That afternoon, the presidential party winged on to St. Louis, where they visited the plant of the McDonnell Aircraft Corporation, whose employees had built the capsules for the Mercury man-in-space flights. Now the firm had a contract to make two-man capsules for Project Gemini, a forerunner of a flight to the moon.[7]

Eight months after President Kennedy returned to the White House from his tour of space installations, the Soviets sent *Vostok VI* into orbit. This mission gained special attention in the West for the pilot was a woman, Valentina Tereshkova. The selection of a woman for the hazardous flight caused some raised eyebrows in the United States, but the Russians thought nothing unusual about their choice. Indeed, Tereshkova performed her task as skillfully as had her male counterparts.

Later Valentina married another Soviet cosmonaut, Adrian Nikolayev, who had made an earlier space flight, and the daughter born to the couple could proclaim that she was the only human being with two cosmonauts for parents.

Senator Edward M. "Ted" Kennedy, the president's youngest brother at age 31, was presiding over a routine debate in the Senate. The chamber was only about half full, and those present were paying little or no attention to a speech by Winston Prouty, a Republican senator from Vermont. Young Kennedy was going through the pretense of listening but was actually signing a stack of letters.

In the Senate press gallery, a media liaison official, Richard Riedel, read an Associated Press teletype:

BULLETIN

DALLAS, NOV. 22, 1963 (AP)—PRESIDENT KENNEDY WAS SHOT TODAY JUST AS HIS MOTORCADE LEFT DOWNTOWN DALLAS. MRS. KENNEDY JUMPED UP AND GRABBED MR. KENNEDY. SHE CRIED, "OH, NO!" THE MOTORCADE SPED ON.

Riedel, who had been a Senate employee for 50 years, ran to the floor where Senator Prouty was droning on about federal library services. Riedel hurried toward the rostrum and Ted Kennedy glanced up in surprise. No employee ever had interrupted a debate by dashing onto the Senate floor.

Young Kennedy looked down at Riedel. Riedel started, "Your brother . . . "

"Yes, what about him?"

"The president. Your brother . . . he's been shot."

At Hickory Hill, the suburban Washington estate of Attorney General Robert F. "Bobby" Kennedy and his wife, Ethel, the telephone by the pool rang at 1:46 P.M. Ethel picked up the receiver and the caller said he was J. Edgar Hoover and that he wanted to speak to the attorney general. Bobby Kennedy was concerned. Only in the direst emergency did the FBI director ever phone him at home.

"I have news for you," Hoover said. "The president's been shot."

Replacing the telephone, Bobby walked back to Ethel with his hand across his mouth, tears welling in his eyes.

A few days later, John Fitzgerald Kennedy, age 45, the visionary who had launched America on an odyssey to the moon, was laid to rest under an eternal flame at Arlington Cemetery on a Virginia height overlooking Washington. A nation mourned.

John Kennedy's untimely death not only cast a pall over Washington and the nation but it also raised grave doubts at NASA over whether the moon-landing program would be continued. Those uncertainties were soon put to rest when Lyndon Johnson, after being sworn into the office of president, pledged that the space project would continue, in his words, "full blast."

Project Mercury was now concluded. It had demonstrated man's proficiency as a pilot in space flights lasting up to 24 hours, each mission carrying one astronaut. However, much testing of spacecraft and of the reactions of persons aboard the vehicle had to be accomplished. So Project Gemini was inaugurated as a transition between Mercury and Project Apollo, the moon-landing missions.

Gemini would involve a series of longer orbital flights of up to two weeks' duration, each spacecraft carrying two astronauts. Gemini also would have to perfect techniques of orbital rendezvous and docking—essential ingredients for a moon landing.

Lifting off late in August 1965, *Gemini 4*, carrying Captain Edward White and Captain James McDivitt, completed 62 earth revolutions. Through the miracle of television, perhaps a fifth of the world's population gawked in fascination as White took a stroll, untethered, outside the capsule. Down below, John Q. Public's overriding question was: what keeps the astronaut from floating away and being lost in space?

Amid the widespread hoopla on the safe return of Jim McDivitt and Ed White, Lyndon Johnson, himself flushed with pride and emotion, threw a monkey wrench into the Pentagon's machinery. Ignoring established procedures, Johnson jubilantly announced that he was promoting the two Air Force officers from major to lieutenant colonel. The president had not bothered to find out that White and McDivitt had only recently been elevated from captain to major. Both astronauts were delighted over their rapid rise up the rank totem pole.

Within hours of his spur-of-the-moment promotions, Johnson began hearing rumblings from the Department of Defense and NASA. So the president—again without consulting the Pentagon—accelerated promotions for Lieutenant Commander John Young and Major Gus Grissom, who had flown *Gemini 3* three months earlier. Now it was Grissom's and Young's turn to be delighted. "At this rate, I'll soon be a four-star general," Grissom quipped to friends.

Five days after their show-stopper, Jim McDivitt and Ed White flew to Washington from Houston with their wives and children at the invitation of President Johnson. No sooner had the helicopter bringing them from nearby Andrews Air Force Base touched down on the White House lawn than Lady Bird Johnson, gushing in her excitement, said she wanted all of the visitors to spend the night. Babysitters would be provided.

In a brief ceremony, the two astronauts heard the president call them "Christopher Columbuses of the 20th Century." Johnson, who was inclined to be impetuous in his public statements at times, declared that the United States had now caught up with the Russians.

After a parade up Pennsylvania Avenue and lunch with Vice President Hubert Humphrey and a galaxy of congressional leaders, McDivitt, White, and their wives were celebrity guests that night at a lavish reception laid on by the State Department. Suddenly, a door was thrown open and in walked Lyndon Johnson.

Those who knew the president best could tell that he was seething. At the Paris Air Show, then in progress, the Russians had humbled the Americans. There, the president had just learned, Yuri Gagarin, the first man to orbit the earth, was standing proudly by his spacecraft, shaking hands with thousands of visitors from all over the world and passing out *Vostok* pins. Worse, Johnson had been told, the French press made it a point to say that crowds were shunning the American pavilion. This was more than a slap at Uncle Sam, the president no doubt felt. This was a Russian slap at *him*.

Now Johnson strode up to Ed White and Jim McDivitt. "I want you to join our delegation at the Paris Air Show," the president declared. Looking startled, one astronaut asked, "When?" Replied Johnson: "Right now—as soon as your wives can pack."

It was not merely an invitation: Lyndon Johnson also was the armed forces commander in chief. Patricia White and Patricia McDivitt were thunderstruck. "But, Mr. President," they said in unison, "We have nothing to wear!"

Never mind, Johnson said, brushing off their protests. "Lady Bird, Lucy and Lynda Bird have closets full of clothes," he said.

A mad dash in a police-escorted limousine to the White House, a quick outfitting of the two Patricias in Lady Bird's bedroom, a rush to board a government jet at Andrews Air Force Base, and a midnight takeoff for Paris. Also on board with the McDivitts and the Whites were Vice President Hubert Humphrey (who had been one of Dwight Eisenhower's most strident space-program critics) and NASA Director James Webb.

The *Gemini 4* men made it to Paris for only the final day and a half of the 11-day show. But they accomplished Lyndon Johnson's purpose. Throngs of people now shunned Yuri Gagarin for the most recent space explorers, Ed White and Jim McDivitt.

Meanwhile, Lieutenant Colonel Frank Borman and Lieutenant Commander James Lovell, in *Gemini 7*, rendezvoused 195 miles above the earth with *Gemini*

*6*, carrying Commander Walter Schirra and Captain Thomas Stafford. This set the stage for an even more intricate maneuver a few weeks later involving *Gemini 8*.[8]

Early in the morning of March 18, 1966, an Agena rocket with a funnel-like docking collar was sent into earth orbit, and an hour and 41 minutes later, Neil A. Armstrong, a civilian, and Major David R. Scott blasted off to chase the Agena across 105,000 miles of space. *Gemini 8*, with Armstrong as the command pilot, overtook the rocket after three orbits and Armstrong successfully guided the nose of the spacecraft into the Agena's docking collar. It was the first manual space-docking maneuver in history.

Thirty minutes after the docking, a jet thruster malfunction sent the coupled vehicles into a dizzying spin. Acting quickly and coolly in the life-or-death situation, Armstrong unlocked the *Gemini 8* from the Agena. However, the spacecraft continued spinning like a top until, after a half-hour, Armstrong succeeded in using 16 reentry rockets to stabilize the vehicle. With what a NASA official called ''extraordinary piloting skill,'' Armstrong brought the *Gemini 8* to an emergency splashdown in the Pacific, ten hours and 42 minutes after lift-off.

Then, in July 1966, Commander John Young and Air Force Captain Michael Collins completed 43 earth revolutions in *Gemini 10*, a flight that proved the feasibility of refueling in space after docking with an Agena rocket already in orbit. Collins became the first astronaut to work outside the spacecraft twice on the same mission.

Progress continued to be made. On September 15, Lieutenant Charles ''Pete'' Gordon, Jr., in *Gemini 11*, set an altitude record of 850 miles, and two months later, Major Erwin E. ''Buzz'' Aldrin, Jr., worked outside *Gemini 12* in space for a record five hours and 30 minutes.

That concluded the Gemini program, which had been a brilliant success: all ten manned flights had accomplished their missions.

At the same time, NASA leaders were puzzled. During the two years that Gemini had been in progress, the Soviets had been orbiting a series of satellites so large that they clearly were intended for a moon shot, but none of them had been manned.

Privately, officials involved in American space flights were offering up praise to their lucky stars. Despite the pioneering nature of the program, not a single astronaut had lost his life in space flight or even had been injured.

Now NASA was preparing for the first step in Project Apollo—launching a three-man crew into earth orbit. Selected for the mission were Lieutenant Colonel Gus Grissom (a veteran of two space flights), Lieutenant Colonel Edwin White (who had been spot-promoted by President Johnson and sent on the midnight ride to Paris), and Navy Commander Roger B. Chaffee, who was eagerly anticipating his first journey above the atmosphere.

These first Apollo astronauts, as with their colleagues, were keenly aware of the perils as trailblazers in space flight. A few months earlier, Gus Grissom told reporters: "If we die, we want people to accept it. We are in a dangerous job, and we hope that if anything happens to us, it will not delay the program. The conquest of space is worth the risk of life."

On the evening of January 27, 1967, Grissom, Ed White, and Roger Chaffee, burdened by bulky spacesuits, were perched in a capsule atop a Saturn 1-B rocket, ready to be blasted skyward in what was described as a "routine exercise." It would be a two-week flight that was entered in NASA books as Apollo 204.

Developed by Wernher von Braun and his Huntsville team, the rocket was 224 feet long. Its maiden suborbital flight from Cape Kennedy (renamed in honor of the late president) had been on February 26, 1966, and, with an unmanned Apollo spacecraft as a payload, performed flawlessly.

Suddenly, at 6:31 P.M., a voice (probably Gus Grissom's) was heard to shout: "Fire in the spacecraft!" Then another loud voice: "Get us out of here!"

Technicians standing on the gantry structure at the level of the spacecraft saw a blinding sheet of flame engulf the cabin, no doubt killing the three astronauts instantly. The capsule was a charred ruin; the Saturn 1-B was virtually undamaged.

America was in shock. Members of Congress took to the floor demanding "answers." Hearings were held. NASA promptly formed a board of inquiry (which included astronaut Frank Borman), but the cause of the disaster was never pinpointed. Many experts felt that the blaze was started by a short-circuit in the spacecraft.[9]

On April 3, 1967, a Grumman Gulfstream was taxiing for a takeoff at Washington National Airport. On board were Bob Gilruth, director of the Manned Spacecraft Center at Houston, and his deputy, George M. Low. The two NASA men had attended a series of meetings in the capital and were about to return to their home base. There was a loud revving of the engine as the pilot prepared to zip down the runway when a cryptic message came in from the tower: the passengers were to return to the terminal and wait in the pilots' lounge.

Less than a half-hour later, nearly the entire NASA high command traipsed into the lounge: Jim Webb; his deputy, Robert C. Seamans, Jr.; George E. Mueller, associate administrator; and Apollo program director Samuel C. Phillips.

Jim Webb wasted no time on chit-chat and got right to the point: Apollo was faltering badly. Since the Apollo 204 tragedy, the program was virtually on hold, and there were only 33 months remaining in the decade to put a man on the moon. Webb instructed George Low to take charge of rebuilding the spacecraft and to make certain that the deadline was met for a lunar flight.

# 23

# A Christmas Eve Spectacular

Aerospace scientists in mid–1967 agreed: the race to land a man on the moon between Uncle Sam and Ivan the Bear was nearing the home stretch, and the competitors were running neck and neck. Whether Uncle Sam would cross the finish line first depended in a large measure on the development of Saturn 5, by far the largest rocket ever built.

Everything about the Saturn 5 was gigantic. Weighing nearly 6 million pounds when fully fueled, the rocket was nearly 200 times as heavy as the Peenemünde V–2 and four times more powerful than any American or Soviet rocket ever launched.

When topped with the Apollo spacecraft, the Saturn 5 stood 363 feet tall, six stories higher than the Statue of Liberty in New York harbor. It was designed to have the capability of boosting a spacecraft weighing nearly 50 tons to the moon or to place a 150-ton payload into orbit around the earth.

The Saturn 5, which had been in development since five months after President Kennedy issued his call for a moon trip back in 1961, had three stages. Its first stage contained five engines, which together generated 7 million pounds of thrust at sea level. This first stage was the biggest aluminum cylinder ever machined. Its valves were as big as beer barrels; its fuel pumps (for feeding engines at the rate of 700 tons of fuel a minute) were larger than refrigerators. Its pipes were big enough for a man to crawl through, and its engines were the size of trucks.

The first stage would boost the upper two stages and the spacecraft on the nose of the third stage to earth-orbital height (nearly 120 miles) and to near-orbital speed of more than 15,000 miles per hour. Finally, the third stage would inject the spacecraft into orbit around the moon.

There had been a mad scramble within the entire aerospace industry to gain the Saturn 5 contract. "But to give the entire financial and technological plum to a single contractor would have made all the others unhappy," Wernher von Braun once recalled. "More important, Saturn 5 needed the very best engineering and management talent the aerospace industry could muster, so by breaking up the parcel into several pieces, more top people could be brought to bear on the project.''[1]

Consequently, the Boeing Company was awarded the contract on the first stage, North American Aviation won the second stage, and McDonnell Aircraft fell heir to the third stage. Systems engineering and overall responsibility for the rocket's development was placed in the hands of von Braun's team at the Marshall Space Flight Center.

When President Kennedy had appointed ex-Marine Jim Webb to be NASA administrator six years earlier, media editorials had criticized the choice of a man who admittedly knew little about space technology. Now the reason for Kennedy's choice became clear: uniquely tight procurement procedures introduced by Webb resulted in acquiring billions of dollars' worth of exotic hardware and facilities without overrunning initial cost estimates and without a hint of ''procurement irregularities''—that is, cheating.

Saturn 5 had become big business. NASA had taken over the cavernous Michoud Ordnance plant—46 acres under one roof—near New Orleans and assigned it to Boeing for production of Saturn's first stage. An area of 13,350 acres in remote Hancock County, Mississippi, on the Gulf of Mexico, was acquired. There huge stands were built for the static testing of Saturn 5's first and second stages.

Shipment of the various stages between Huntsville, Michoud, the Mississippi test center, two California contractors, and Cape Kennedy had evolved into a major logistical operation. Soon the Marshall Space Flight Center was running a small fleet that included the barges *Promise*, *Orion*, and *Palaemon*; for carrying cargo through the Panama Canal, there were the ocean-going ships *Steel Executive* and *Point Barrow*.

When rapid shipments were required, the Huntsville center had two converted Stratocruisers, aptly named *Super Guppy* and the *Pregnant Guppy*. Bulbous *Super Guppy* was the only airplane in the world capable of carrying a complete Saturn stage and, when loaded, flew smoothly at a 250 miles per hour cruising speed.[2]

Through the use of computer analysis, vibration tests, simulations of the space environment, and ground firings, nothing about the Saturn 5 was left to chance. An automatic electronic checkout system monitored thousands of points on the rocket right up to the moment of lift-off.

''It's not like the early days,'' Wernher von Braun told the press. ''Then it was a kind of trial and error method. You designed these things, but didn't really know what they would do until you pushed the button.''[3]

The Apollo spacecraft had three basic components: the command module (CM), the service module (SM), and the lunar module (LM). Built by North American Aviation, the CM was the control center and the three-man crew's basic working and living area. The SM provided propulsions, power, and storage room for consumables, while the LM would ferry two of the three astronauts from the orbiting CM to the moon.

One of history's most complicated contraptions, the lunar module, although only a part of the Apollo spacecraft, was a huge vehicle in its own right. It was 12 feet in height, 14 feet in circumference, and almost 17 tons in weight.

Prime contractor for the LM was the Grumman Aircraft Engineering Corporation, which occupied an 850-acre complex at Bethpage, Long Island. The LM was an enormously sophisticated piece of equipment, and on its perfection (or lack of perfection) would rest the lives of at least two of the moon astronauts. Designing the LM's electrical power and control systems was an especially painstaking and intricate procedure.

In the meantime, a sophisticated facility for testing the LM under the simulated conditions expected to be encountered in the lunar-landing mission was designed by Burns and Roe (as a subcontractor to Grumman). The plans were approved by NASA and the LM propulsion test facility was built at White Sands, New Mexico.

The actual LM that would land on the moon was brought to White Sands to undergo exhaustive trials in simulated climatic conditions thought to be like those around and on the moon.

There were two stages to the LM: descent and ascent. With its four bent landing legs, each containing aluminum shock absorbers to lessen the impact, and the other protrusions, the LM had the appearance of a gigantic insect that Hollywood might have depicted as being alien life from some distant galaxy.

Meanwhile at the Kennedy Space Center, the Vehicle Assembly Building (VAB) was constructed to provide a place for erecting the huge Saturn 5 and its spacecraft. Reputed to be the world's largest building, the VAB was as high as a 54-story structure and almost twice the size of the sprawling Pentagon. Kennedy's 26,132 technicians joked that clouds formed and it rained inside the cavernous VAB.

Assembling the three stages of the Saturn 5 and its spacecraft required special skills and steady hands. A story making the rounds had it that the crane operator who gingerly set the 88,000-pound second stage on top of the first stage had to qualify for the job by lowering an identical weight until it touched a raw egg—without cracking the shell.

In the wake of the Apollo 204 disaster that snuffed out the lives of Gus Grissom, Edwin White, and Roger Chaffee, those directing the moon-landing program were engaged in an urgent battle against time. "The work to be done [to meet President Kennedy's target date] appeared to be overwhelming," recalled George Low, deputy director at Houston's Manned Spacecraft Center. "It dictated 18-hour days, seven days a week. Our briefcases were our offices, our suitcases our homes."

Intense scrutiny was given to the spacecraft in which the three astronauts had met their deaths. George Low formed a Configuration Control Board, which met at noon each Monday and continued far into the night. During the next two years, the board, often in heated discussions, considered 1,697 changes in the Apollo spacecraft and approved 1,341 of them.

For weeks, rumors reached NASA leaders that the Russians were about ready to try a unique first: two spacecraft exchanging crews with one another while in orbit. At 3:36 A.M. (Moscow time) on April 23, 1967, *Soyuz 1*, a spacecraft thought by American intelligence to have room for four cosmonauts, lifted off from the launch port at Kazakhstan in the Republic of Baikonur. Despite its roominess, Colonel Vladimir M. Komarov, who had been the pilot in a 1963 space flight, was the only one on board *Soyuz 1*.

After orbiting for about 24 hours in what NASA thought might have been part of the crew-switching operation, a decision was reached on the ground for Komarov to land. Unlike America's flights, in which the capsules splashed into the ocean, the Soviets landed their own on land by means of gigantic parachutes.

Komarov released the chute at an altitude of about 23,000 feet, but the lines snarled and the craft plummeted to the ground, killing Colonel Komarov. Typically, the Kremlin kept the *Soyuz 1* failure and the cosmonaut's death a secret from the Russian people.

By September 1967, the Saturn 5 was ready for a crucial test flight. Should the rocket fail, America's hopes for reaching the moon might perish with it. At the Vehicle Assembly Building, the 465-foot-high door was opened and the upright rocket was rolled into the morning sun and loaded onto a mobile launcher. This entire cargo, in turn, was placed onto a mammoth Rube Goldberg-like crawler. Built by the Marion Power Shovel Company, the crawler had been conceived out of necessity as a means of transporting the Saturn 5 from the VAB to the launchpad, three and a half miles away.

Shipped from Ohio to Cape Kennedy in sections, the crawler was 114 feet wide and had eight tracks, each seven by 41 feet, with cleats like a Sherman tank, except that each cleat weighed a ton. Mounted over these eight tracks was a platform, larger than a baseball diamond.

Perched majestically on the platform of the crawler, the Saturn 5 assembly began inching toward the launchpad at one mile per hour "speed." Enclosed in a glass and metal cockpit, the driver, a Ph.D. clad in a three-piece business suit, wore a mandatory seat belt and was like the helmsman of a ship. Ten hours and 15 minutes after the glacier-like trek began, the crawler reached its destination.

There a tedious and painstaking checkout began. So complicated was the Saturn 5 that the number of printed pages, including interface control documents, needed to do the job numbered in excess of 30,000. There were so many copies of the documents that it would have required a boxcar to hold them.

It took nearly two months to complete checking out the rocket. Then, at 7:01

A.M. on November 9, 1967, the Saturn 5 soared into the heavens on its maiden voyage. Each of the stages performed flawlessly, and the third one boosted the Apollo capsule to 25,000 miles per hour, matching the velocity it would reach on a return trip from the moon. The orbital weight of 278,692 pounds broke all records—and dwarfed the 30-pound satellite that Wernher von Braun and his Huntsville team had sent into earth orbit nine years earlier.

Eight hours and 31 minutes after blast-off, the unmanned Apollo command module splashed into the Pacific Ocean 600 miles off Hawaii—right on target.

In November 1968, the Soviets barged back into the lead in the space race by sending the unmanned capsule *Zond* into orbit around the moon. Apollo leaders were jolted by the feat and Radio Moscow trumpeted: "This is the trailblazer for a manned moon landing."

Five weeks later, Uncle Sam surged out in front in the moon sweepstakes when *Apollo 8*, with Lieutenant Colonel Frank Borman, Colonel William Anders, and Navy Captain James Lovell aboard, orbited the moon ten times on Christmas Eve and Christmas.

A half-billion people saw on television what man had never seen before: the moon, close up. "The vast loneliness is awe inspiring," they heard Jim Lovell exclaim. On Christmas Eve, Bill Anders commented: "For all the people on earth, the crew of *Apollo 8* has a message we would like to send you." Pausing for a moment, he began reading: "In the beginning God created the Heaven and the Earth." After four verses of Genesis, Jim Lovell took up the reading: "And God called the light Day, and the darkness he called Night." At the end of the eighth verse Frank Borman picked up the words: "And God said, let the waters under the Heavens be gathered."

It was a time of rare emotion—in the orbiting *Apollo 8*, in Huntsville, in Houston, in Washington, at Cape Kennedy, and throughout much of the world. The mixture of the Christmas season, the immortal words, the ancient and inscrutable moon, and the dazzling new technology combined to create a scenario never known before by humankind.

On January 9, 1969, in Washington, the first circumnavigators of the moon officially were installed in the select ranks of national heroes: a gold medal and praise as "history's boldest explorers" from President Lyndon Johnson, rousing cheers from throngs along Pennsylvania Avenue, and a prolonged standing ovation from a joint session of Congress.

As a boy growing up in Gary, Indiana, Borman was sickly, plagued with sinus, mastoid, tonsil, and adenoid troubles. When he was five, his family moved to a warmer climate in Tuscon, Arizona, at the suggestion of doctors. Young Frank steadily grew stronger, his ills evaporated, and he became quarterback on the high school football team that won the state championship.

A natural leader with an intense competitive spirit, Borman graduated eighth in a class of 670 at the United States Military Academy and earned his wings with the Air Force in 1951. Outgoing and witty, Frank Borman, after *Apollo 8*, was described in the *Toronto Globe and Mail* as "looking more like a successful automobile salesman than an heroic astronaut."[4]

In the Capitol, 40-year-old Frank Borman, as spokesman for the *Apollo 8* crew, stood at the rostrum where such American legends as Charles A. Lindbergh, General Dwight Eisenhower, and General Douglas MacArthur had addressed the legislative body. In his speech, Colonel Borman received his loudest applause, mixed with ringing laughter, when he described the crew's reading from the book of Genesis on Christmas Eve. "One of the things that was truly historic," Borman declared with a straight face, "was that we got that good Roman Catholic, Bill Anders, to read from the King James version," a Protestant translation.

Then, looking down at the black-robed justices of the Supreme Court, which had ruled against reading the Bible in public schools, the *Apollo 8* commander smiled and added: "But now I see the gentlemen in the front row, I'm not sure we should have read the Bible at all."[5]

Frank Borman's good-natured quip proved to be prophetic. A week later, a self-proclaimed atheist in Texas filed suit against NASA, demanding that future astronauts be prohibited from reading the Bible while in "government property"—meaning an orbiting spacecraft.[6]

On the same day that Colonel Frank Borman jokingly admonished the Supreme Court justices, Robert Gilruth, director of the Manned Spacecraft Center in Houston, ended weeks of rife speculation over which of the astronauts would be selected for *Apollo 11*, the highly coveted moon-landing mission. Gilruth named three veterans of earlier space flights: civilian Neil Armstrong, Colonel Edwin "Buzz" Aldrin, and Lieutenant Colonel Michael Collins.

Frank Borman had seemed to be the odds-on choice to be the first man to set foot on the moon. But after *Apollo 8*, in deference to his wife and children, he decided to make no more space flights. Borman's backup was Armstrong, so, in the normal rotation used by NASA, he would command the *Apollo 11* flight.[7]

Deke Slayton, who had been grounded by doctors years earlier and since had become a power within NASA as director of flight crew operations, decided that Neil Armstrong would get out of the lunar module first when its descent stage reached the moon, thereby cementing his place in history. "I figured the commander ought to be the first guy out," Slayton explained to newsmen. Did Armstrong pull his rank, as was widely conjectured? No way, Slayton replied firmly. "I was never asked my opinion," said Armstrong.

The target date for blasting off to the moon was set for Wednesday, July 15, 1969. At a news conference, James Webb and a galaxy of top NASA leaders groped in their engineering lexicon for phrases to capture the significance of man's first landing on the moon. Words failed all of them save one—Wernher von Braun.

To the German-born rocket pioneer, the moon flight was nothing less than a step in human evolution comparable to the time when life on earth emerged from the sea and established itself on land. "We are now going to establish mankind in space," he explained.

In order to reach the threshold of one of history's mightiest endeavors, Project Apollo had developed into an organization rivaling in size and scope the super-secret Manhattan Project that created the A-bomb in World War II. There were now 34,126 NASA employees, 100,000 scientists and technicians, and 377,032 contractors, subcontractors, and vendors involved.

At 12:30 P.M. on May 20, 1969, the 456-foot hangar doors in the Cape's Vehicle Assembly Building were opened and the mighty Saturn 5 was rolled out into the bright Florida sunshine. It was one month and 26 days before the rocket was to boost three Americans to the moon.

# 24

# "The *Eagle* Has Landed!"

> It was the first time in the history of the world that explorers went to a new land without weapons of any sort.
>
> —Neil A. Armstrong, August 23, 1969

Lyndon Baines Johnson arrived early and virtually unnoticed among the throngs that had gathered at Cape Kennedy on July 17, 1969, for an eyewitness view of what could be one of humankind's most spectacular feats. Within minutes, Johnson was fuming. He had accepted President Richard Nixon's invitation to attend the launching—his first public appearance since leaving office the previous January—and now he had been shunted off to the "peanut gallery," as he would call the bleachers.

Once the world's most powerful man, Johnson found this cavalier treatment quite galling. Even after retiring to his ranch in Texas, he had taken with him the lingering trappings of the presidency: the grant of $375,000 annually, the office space, free mailing privileges, the military helicopter, the Secret Service men. But the real power had vanished like wisps of smoke in a hurricane.

A few people strolled over to shake Johnson's hand as he sat in the scorching sun, a pitiful figure. Most of the crowd, however, had binoculars trained on the waiting spacecraft that would take the first humans to the moon.

"My trousers stuck to the back of my legs, the sweat from my hair kept dripping down my neck," Johnson remembered years later. "My stomach was upset. I knew right then that I shouldn't have come. I didn't want to go in the first place, but I just didn't feel right saying no to the President's invitation.

"I hated being there. I hated shaking hands with all those people, pretending I remembered who they were when I'd never seen them before in my life. Each conversation was like a goddamned quiz. I hated every minute of it."[1]

Before dawn that day, astronauts Neil Armstrong, Edwin "Buzz" Aldrin, and Michael Collins were awakened in their quarters at the Operations and Checkout Building in the Kennedy Space Center. After breakfast and a last-minute physical examination by doctors, the trailblazers donned spacesuits and, burdened by their heavy gear, waddled toward the transfer van that would carry them to launchpad 39-A.

There stood the awesome Saturn 5 rocket, the early morning rays of a bright sun glistening on its smooth surface. As the three astronauts rode in the elevator the 360 feet to where the Apollo spacecraft sat perched on the tip of the rocket, none revealed outward signs of undue concern for their safety. But seeping into their beings, perhaps, were the prophetic words of their colleague, Gus Grissom, a few weeks prior to his fiery death: "If we die, we want people to accept it. For the conquest of space is worth the risk of life."

If Buzz Aldrin and Michael Collins ever had harbored resentment over the fact that Neil Armstrong had been selected to be the first human to set foot on the moon, they disguised it well. Collins remained his customary happy-go-lucky self and Aldrin had told the media, half-convincingly, "It's all right with me if Neil is first."

Many years later, Neil Armstrong recalled his thoughts as the moon pioneers headed for the Apollo spacecraft:

> Although confident, we were certainly not overconfident. In research and exploration, the unexpected is always expected. We would not have been surprised if a malfunction or an unforeseen occurrence prevented a successful moon landing.
>
> We knew that hundreds of thousands of Americans had given their best. Now it was time for us to give our best.

Blond, blue-eyed, and a sturdy 165 pounds, the boyish-looking five-foot, 11-inch Neil Armstrong was born on a farm near Wapakoneta, Ohio, on August 5, 1930. Although his parents were extremely devout, he was the only one of the 19 Apollo astronauts who professed no religious affiliation.

Armstrong had his first ride in an airplane, a Ford trimotor job, when he was six, and was hooked on aviation. Working afternoons to pay for flying lessons in his early teens, Armstrong took to flying like a bird dog to pheasants. On the day he turned 16, he received a pilot's license. As a teenager, Armstrong was a zealous Boy Scout and played baritone horn in a jazz combo (eventually he mastered four musical instruments, including the piano).

After spending two years at Purdue University, Armstrong was called to active duty by the Navy, earned his wings, and flew 78 combat missions during the Korean War, being shot down on one flight. Returning to civilian life, he earned a degree at Purdue and, in 1955, became a test pilot, logging more than 1,100 hours. His flights in the revolutionary X–15 rocket plane were at speeds up to 4,000 miles per hour and altitudes of 40 miles.

"Neil flies an airplane like he's wearing it," a friend once remarked.

After seven years as a test pilot, Armstrong volunteered for the astronaut program and was accepted in September 1962.

Armstrong had a sense of humor so dry it sometimes escaped the inattentive listener. Generally laconic, he rarely became eloquent except when speaking of aeronautics. His only bad habit was smoking cigars—one per month.

Unlike other American astronauts, Armstrong shunned calisthenics and sports. He explained, "Every human being has a finite number of heartbeats available to him, and I don't intend to waste any of mine running around doing exercises."

Now, on Moon Day, Neil Armstrong and Michael Collins were being strapped into their seats and connected to the spacecraft's complicated life-support system. Buzz Aldrin waited near the elevator a floor below for his turn to climb into the capsule. Gazing around at the vast panorama, Aldrin could see people and cars jamming the highways and beaches to the limits of his vision.[2]

Colonel Edwin Aldrin, Jr., would be, if all went well, the second man to set foot on the moon. Born in Glen Ridge, New Jersey, on January 20, 1930, he was a unique combination of skilled pilot and Ph.D., having earned his doctorate from the Massachusetts Institute of Technology. After graduating from the United States Military Academy third in his class in 1951, Buzz Aldrin flew 66 combat missions in Korea and was selected as an astronaut in 1963.

Perched atop a huge rocket waiting to be blasted into space (or into eternity) was nothing new for the three veteran astronauts. But Michael Collins, while heavily involved in a final checkup of systems in the command module (named *Columbia*), reflected that the mission had only a 50–50 chance of succeeding. There were thousands of things that could go wrong—including human error by the crew. In his mind's eye, sitting there ready to bolt into outer space, Collins envisioned the next day's newspaper headline: "MOONSHOT FALLS INTO OCEAN." Then the second head: "Mistake by Crew. Last Transmission from Armstrong Prior to Leaving the Pad Was 'Oops!' "

Lieutenant Colonel Michael Collins was born on October 31, 1930, in Rome, Italy, where his father, a two-star general, was military attaché. Michael's brother, 13 years his senior, was a brigadier general, and his uncle was General J. Lawton Collins, a World War II hero and later Army chief of staff.

As an "Army brat," Michael Collins grew up at six military posts. While attending St. Albans, an Episcopalian prep school in Washington, D.C., he was a guard on the football team, shortstop on the baseball team, and captain of the wrestling team. A teacher at St. Albans, Ferdinand E. Ruge, recalled, "Mike wasn't the romantic or cavalier type you would expect from an astronaut. He was a very quiet boy—very self-possessed and imperturbable. However, there was something puckish about Mike." Professor Ruge added, "Beneath that quiet exterior you always wondered what mischief he was about."

In 1952, Collins graduated from West Point and, like Buzz Aldrin, opted to take a commission in the Air Force. In 1960, he was accepted at the Air Force's test-pilot school at Edwards Air Force Base in California, and two years later he became an astronaut.

A wiry man, five feet, 10 inches tall and weighing about 165 pounds, Mike

Collins was modest, had a keen sense of humor, and claimed to have a passion for anonymity.

While Neil Armstrong, Buzz Aldrin, and Michael Collins waited for the countdown to end, a short distance away three helicopters touched down near the bleachers where Lyndon Johnson and a bevy of senators, congressmen, cabinet members, and assorted VIPs were gathered. As a blanket of dust whipped up by the choppers' blades drifted over the bleachers, a band struck up *Hail to the Chief*.

However, it was not President Richard Nixon who ascended from one helicopter, but rather his vice president, Spiro Agnew, who had been an obscure Maryland politician until Nixon tapped him to be his running mate a year earlier. Agnew was given the red carpet treatment, with a proper amount of bowing and scraping by a bevy of nabobs. Lyndon Johnson, lost in the crowd and covered with a cake of perspiration and dust, fumed some more.[3]

Since being inched from the Vehicle Assembly Building to launchpad 39-A 57 days earlier, Saturn 5 had been undergoing the most exhaustive checkout that ingenious scientific minds could conceive. There were some 500 engineers and technicians, working 12- or 13-hour days in the pre-lift-off operation.

Rocco Petrone, who had succeeded the pioneer Peenemünde expert Kurt Debus as director of launching operations, recalled the tension-racked period leading up to the lift-off:

> The pressure on our people was pretty severe. At a launch, a person just sat there glued to his console, watching the needles for any sudden changes, knowing that he would be committing this big vehicle, with men aboard. The entire world was watching.
>
> I walked through the console panel area right up to about the last 45 minutes before lift-off. I'd be checking on alertness, especially among men who had been working long hours. Were they fatigued? Were they concentrating on the dials? Was there any unnecessary chit-chat going on? . . . The team had to be as well rehearsed as any ballet, or any football team.

Now the historic flight of *Apollo 11* was ready to begin. At 9:32 A.M. (Eastern Standard Time), the astronauts were perched atop a pillar of fire as the first-stage engines of the Saturn 5 ignited for launch. Soon, the entire rocket-spacecraft assembly roared skyward, at first ponderously, then with greater and greater velocity. Thrown left and right against their straps in fitful jerks, the three astronauts felt as though they were riding a wild beast and hoped the animal knew where it was going.[4]

Twelve minutes after blasting off, *Apollo 11* was in orbit around the earth. As it made one and a half revolutions, the astronauts, along with Mission Control in Houston, checked all systems. Ground stations in the global network tracked the spacecraft to determine if it were following the proper trajectory. Until this point, Mission Control could have altered the flight plan, causing the vehicle to remain in earth orbit or preparing it for an emergency landing.

Two hours and 42 minutes after leaving earth, Houston gave the green light, and the vehicle broke from the earth's gravitational pull (by increasing velocity to in excess of 25,000 miles per hour) and began the 238,857-mile, 73-hour trip to the moon.

On the fourth day of the flight, with the pockmarked moon as a spectacular backdrop, the three crewmen began the critical maneuver of inserting the spacecraft into orbit around the moon. All were aware that the honeymoon was over. As Michael Collins would express it: "We were about to lay our little pink bodies on the line."[5]

Lunar orbit was achieved by slowing the spacecraft's velocity enough to allow it to be captured by the pull of the moon's gravity. After circling the planet ten times, the three men were awakened at 7:00 A.M. on the fifth day. Struggling into bulky spacesuits in the cramped *Columbia*, Neil Armstrong and Buzz Aldrin transferred to the lunar module, which they had christened *Eagle*. This was the moment of truth in America's decade-long endeavor to put men on the moon.

Seven months earlier, in *Apollo 10*, Colonel Thomas P. Stafford, Commander Eugene Cernan, and Commander John Young had performed a dress rehearsal of the mission by orbiting the moon, but no one had descended to the lunar surface in the weird-looking ferry. While the descent maneuver had been simulated at the White Sands testing station, no one knew for sure if all systems would work in the true moon environment.

Should the electrical and control systems and the sophisticated computer, which were designed by some of America's most brilliant scientists and engineers for the *Eagle*, prove to be flawed, the moon ferry might go awry, taking Buzz Aldrin and Neil Armstrong to their deaths.

Now Armstrong and Aldrin crawled out of the *Columbia* and squeezed into the *Eagle*, where they stood, elbow to elbow, at the controls. Collins disconnected the electrical umbilical cord that connected *Columbia* and *Eagle* and as his two companions began their descent, he called out on the radio: "You cats take it easy on the moon!"[6]

*Eagle*'s control system received range data from a radar for the main guidance computations during the descent. An environmental control system regulated cabin conditions, and zinc batteries helped to provide electrical power for communicating by radio directly with Mission Control in Houston and with Michael Collins in the orbiting *Columbia*.

Collins' movie camera was grinding furiously, and he heard Neil Armstrong exult over the radio: "The *Eagle* has wings!" Collins thought that the outlandish contraption, with its four extended legs, did not look like any eagle he had ever seen.

At 3:08 P.M., when *Eagle* was behind the moon and out of touch with Houston, Armstrong and Aldrin made ready for the final, precarious run to the lunar surface. Tension built at Mission Control until 3:46 P.M., when radio contact was reestablished with Collins in the orbiting *Columbia*. Collins reported to Houston: "Things are going just swimmingly, just beautiful!"

At Mission Control, after the 38-minute radio blackout, Flight Director Christopher Kraft, Jr., joined scores of others in issuing deep sighs of relief. Over the years, Kraft's voice had become known to hundreds of millions of television viewers as he calmly directed the Mercury, Gemini, and Apollo flights. His coolness and ability to make quick, accurate decisions in emergencies had inspired the confidence of the astronauts and assured their safety.

Christopher Kraft, Jr., 45 years old, had hoped to be a major league baseball player, and he batted .325 for the Virginia Polytechnical Institute team. As a boy, his two most coveted possessions were baseballs autographed by Babe Ruth and Lou Gehrig. However, the combination of a hand injury and his brilliance in mathematics steered him to a career as an aeronautical engineer.

Within moments of Collins' report that "things are going . . . beautifully," when *Eagle* was 6,000 feet above the lunar surface, a yellow caution light flashed on in the moon-landing vehicle and bells began ringing at Mission Control. Chris Kraft heard Neil Armstrong's firm but calm voice: "Program alarm! It's a 1202!"

In the Apollo checklist, 1202 was an "executive overflow," meaning that the *Eagle*'s computer had been called on to do too many things at once and was forced to postpone some of them.

Those at Mission Control masked the near-panic that gripped them. Chris Kraft, outwardly calm and collected, felt an especial surge of concern: he had been responsible for the software in the *Eagle* computer, the logic that made it all work. What had gone wrong? Was the Apollo project, so close to a monumental achievement, doomed to failure?

Hundreds of eyes at Mission Control focused on 27-year-old Stephen C. Bales who, despite his relative youth, had been assigned the critical task of assessing any computer problems that might arise. The situation demanded an immediate response. Should Bales make the wrong decision, he might be sending Neil Armstrong and Buzz Aldrin to their deaths.

Swiftly, but showing no evidence of the butterflies that were flapping their wings furiously in his stomach, Steve Bales did a few rapid calculations to determine if *Eagle*'s descent should continue or if the moon landing should be aborted. Pressed for an answer, Steve Bales did not hesitate, calling out in a loud voice: "Go!"[7]

Armstrong and Aldrin continued their descent to an altitude of 3,000 feet and saw that the computer guidance system was taking *Eagle* toward a crater filled with boulders that could wreck the vehicle—and strand them on the moon with no hope of rescue. In this latest crisis, Armstrong took over manual control of the steering. Moments later, a red warning light came on: there were only 115 seconds of fuel remaining.

The *Eagle* touched down on a level plain near the southwestern shore of the arid Sea of Tranquility. Buzz Aldrin felt goose bumps over the strange sensation of actually being on the moon. It had been a close-run thing: only 45 seconds of fuel was in the tank.

At 4:18 P.M.—102 hours, 45 minutes, and 39 seconds after leaving earth—Neil Armstrong radioed to Mission Control: ''The *Eagle* Has Landed!''

''You can be sure there are smiles on faces here!'' Mission Control told the first moon men.[8]

Also in Mission Control, tears welled in the eyes of Wernher von Braun, under whose direction the powerful Saturn 5 rocket had been developed. Since boyhood he had spoken eagerly of man one day exploring the universe and landing on the moon. Beginning as a 17-year-old boy at the Raketenflugplatz outside Berlin, he had built ever larger rockets for that purpose. His intense absorption in creating means for peaceful space exploration had landed him in a Gestapo jail and on the verge of execution. Now his boyhood dreams, his lifelong hopes, had been fulfilled.

Yet, like the others at Mission Control, von Braun's elation was tempered by the specter of the unknown hazards that might still be facing Armstrong, Aldrin, and Collins on their return to earth.

For the next six and one-half hours, Armstrong and Aldrin remained aboard the *Eagle* and kept busy making the adjustments necessary to prepare for ascent back to the circling *Columbia* in an emergency. Perhaps they silently pondered what fate awaited them once they climbed out of the craft and onto the moon. A few experts, prior to the flight, predicted that a good bit of difficulty would be encountered by humans due to the strange atmospheric and gravitational characteristics of the moon.

Their preparations completed, the astronauts opened the *Eagle*'s hatch and Armstrong began backing his way down the ladder. At 1:56 A.M., the first human footprint was planted on the moon.

Armstrong's movements were televised to an estimated one billion awe-stricken earthlings by means of a small camera attached to the *Eagle*. They heard him say: ''That's one small step for man, one giant leap for mankind!''

Fourteen minutes later, Buzz Aldrin came down the ladder. Like any typical moon tourist, Neil Armstrong had his camera ready to photograph his companion's arrival.

Armstrong and Aldrin were in high spirits, hopping about like bunny rabbits to display the ease of movement possible in lunar gravity. With his big backpack and heavy spacesuit, each astronaut weighed about 360 pounds on earth and only 60 pounds on the moon.

There was much serious business to do and little time in which to do it. While the two astronauts set up a seismometer (an instrument that records ground vibrations), collected 45 pounds of rock and soil to take back to earth for analysis, and conducted experiments in maneuverability, Houston came on the air to announce that President Nixon wanted to talk to them. In typical low-key fashion, as though a president talks to men on the moon every day, Armstrong replied: "That would be an honor."

It was a curious situation. Nixon, a 56-year-old native of Yorba Linda, California, had lost to John Kennedy in a bitter and hard-fought 1960 presidential campaign. Had Kennedy been defeated, would two Americans now be standing on the moon?

In the White House, the president, veteran astronaut Frank Borman, and Nixon's aide Robert Haldeman, had been glued to a television set all evening. At the appointed time, Nixon went into the Oval Office next door where television cameras had been set up for a split-screen telephone call to the moon.

As Armstrong and Aldrin paused in their labors to listen, President Nixon, in a tone experienced speechmakers use to convey inspiration, began talking: "Neil and Buzz, because of what you have done, the heavens have become a part of man's world. As you talk to us from the Sea of Tranquility, it inspires us to redouble our efforts to bring peace and tranquility to earth."

My God, Michael Collins reflected in *Columbia*, I never thought of bringing peace and tranquility to anyone! As far as Collins was concerned, the moon voyage was fraught with perils—and that was as far as his heavy thinking had reached.

Once President Nixon signed off and turned his focus back to the bloody war raging in South Vietnam, Armstrong and Aldrin set up an American flag and implanted a plaque that would proclaim to future lunar tourists:

One fifth of the earth's peoples watched the dramatic and emotional flourish to man's conquest of the moon by way of television. In a similar scenario 437 years earlier, a brash young explorer named Christopher Columbus, who had sailed into the unknown from Europe, reached the uncharted New World and implanted the banner of Spain on a small island in the Bahamas. In contrast to the 1 billion people who watched Armstrong and Aldrin, Columbus had an "audience" of three or four bewildered natives.[9]

Now the two moon men climbed into the ascent stage of the *Eagle* to return to *Columbia*. They removed their boots and cumbersome backpacks (which had sustained their life on the lunar surface), opened the hatch and pitched out these items. The descent stage of the vehicle would be left on the moon forever—or possibly until some unknown creatures from another planet arrived to haul it away.

This was a crucial moment in the *Apollo 11* mission. Should the ascent stage engine fail to function, Armstrong and Aldrin would be doomed—marooned on a bleak and inhospitable planet 240,000 miles from their home base, earth.

Tension in the cabin and in Mission Control was thick enough to cut with a knife. Many millions of television viewers were also gripped by jangled nerves, although few realized the full consequence should the engine fail to start.

Suddenly, the ascent-stage engine began to fire. Then the vehicle lifted off the surface, ending the astronauts' stay of 22 hours and 17 minutes. It was 1:55 P.M. About four hours later, the ascent stage and the orbiting *Columbia* docked. Aldrin and Armstrong, one after the other, crawled into the *Columbia*, where a beaming Michael Collins grabbed each man by the hand and pumped it vigorously. Elated like mischievous schoolboys who had just pulled off a unique prank, the three astronauts broke out with a severe case of the giggles.

Now began the 64-hour journey back to earth. At 12:51 P.M. on July 24, 1969, the three Americans splashed down in the Pacific Ocean 950 miles southwest of Hawaii. A helicopter-borne recovery team plucked them from the water and the chopper headed for the aircraft carrier USS *Hornet*.

On the flight deck awaiting the moon men's arrival was President Nixon, who had winged nearly 7,000 miles from Washington on the first leg of a world trip in order personally to greet America's newest legends. Because the command module was named *Columbia*, the president requested the Navy band to play *Columbia, the Gem of the Ocean* when the astronauts stepped onto the deck.

Beaming broadly and freshly shaven, Neil Armstrong, Buzz Aldrin, and Michael Collins, emerged from the chopper; the band broke into the designated tune; and a few thousand sailors stirred up waves with rousing cheers. The tumultuous reception was short-lived: the three astronauts, on the slight chance that they might carry moon germs, were hustled into a mobile quarantine van. "We learned how monkeys in a cage feel," Collins would quip.

President Nixon approached the van and could not disguise his awe and enthusiasm about conversing with men on the other side of the glass who had just returned from the moon. In a fit of buoyancy, Nixon exclaimed: ''This is the greatest week in the history of the world since the Creation!''

Indeed it was an incredible feat, a monumental endeavor. But a few days later, back at the White House, the president was telling Billy Graham of his remark to the moon men. The renowned evangelist pondered for a moment, then replied: ''Mr. President, I know exactly how you felt, and I understand exactly what you meant. But, even so, I think you may have been a *little* excessive!''[10]

The quarantine van holding Neil Armstrong, Buzz Aldrin, and Michael Collins was unloaded from the *Hornet* at Pearl Harbor, Hawaii, and then flown to the Manned Spacecraft Center in Houston. There the astronauts and their precious cargo of moon rocks and soil were transferred into an 83,000-square-foot lunar receiving laboratory.

After 18 days ''in custody'' (as Michael Collins described it), the three moon men were given conquering heroes' parades in New York City, Chicago, and Los Angeles. Then, flying in the presidential jet, *Air Force One*, they went on a 45-day, 24-nation goodwill tour that focused the world spotlight on America's superiority in science and technology. In the Cold War that was raging globally, it was a victory march.

# Epilogue

After Neil Armstrong and Buzz Aldrin returned from the moon, President Richard Nixon was urged by some of his advisors to halt the Apollo program—a political expedient that would avoid the risk of future failure in the extremely dangerous and volatile space-flight business. But Nixon accepted the potential for disaster on the premise that humankind would benefit by continued peaceful exploration of the moon and the universe.

Consequently, during the next two and a half years, six more Apollo flights were launched; not only astronauts, but also geologists and scientists, roamed the lunar surface in the rugged Fra Mauro highlands, the Hadley Apennine areas, the Taurus-Littrow region, and the Descartes highlands. Instead of walking, however, they rode in strange-looking vehicles called lunar rovers, which had been developed at the George C. Marshall Space Flight Center under the direction of Wernher von Braun.

America's triumphant journey to the moon was hailed as the most spectacularly successful exploration in history, the greatest engineering-scientific program of all time, the boldest, most challenging, complicated, and profound endeavor in history. Yet Uncle Sam's colossal achievement might never have happened had it not been for the ingenuity and resourcefulness of the handful of young American Army officers and scientists who, acting largely on their own volition, conceived and implemented Operation Overcast at the conclusion of World War II in Europe.

Had these Army officers and scientists, most of whom were in their late 20s and early 30s, failed in their mission, it may well have been the flag of Communist Russia planted on the moon instead of the Stars and Stripes of free enterprise America.

Of perhaps even greater significance, the Peenemünde team, brought to the United States in Overcast and working with hundreds of American scientists and engineers, closed the 20-year-gap in space technology and developed forerunners of the full range of Pershing, Cruise, and SS–20 series of missiles, which formed the backbone of the NATO armory and deterred the Soviets in Europe until the Soviet Union collapsed in economic chaos.

Likewise, U.S. intercontinental ballistic missiles (ICBMs), antecedents of the V–2 and based in America, provided a deterrent to attack by Russian ICBMs tipped with nuclear warheads and launched from the Soviet Union.

All three of the U.S. 3rd Armored Division officers who discovered the V–2 rocket underground manufacturing plant in the Harz Mountains in April 1945 retired years later as colonels. In 1991, John Welborn was living in Portland, Maine; William Castille was residing in Austin, Texas; and William Lovelady was living in Altamonte Springs, Florida.

Colonel Gervais Trichel, the Army rocket chief in the Pentagon during World War II, who had set Operation Overcast in motion even before it had a code name, retired from the service in 1947. Later, he became manager of the European office of the Chrysler Corporation Defense Operations Division and after that a Chrysler executive in Detroit.

Major James Hamill, who, along with Major William Bromley, ''liberated'' the 100 V–2s from under the noses of the advancing Soviets, retired from the Army in 1961 as a colonel in command of the Ballistic Research Laboratory at Aberdeen Proving Ground.

Major Robert Staver, who, together with 32-year-old Richard W. Porter of General Electric, played a key role in recruiting German rocket experts, left the Army as a lieutenant colonel in December 1945 and joined a California firm in no way connected with rockets. Porter remained with General Electric for many years before retiring in Connecticut to become a consultant to industry.

Colonel Holger ''Ludy'' Toftoy, who had been a prime mover in Operation Overcast and later became not only the leader but the friend and confidante of the Peenemünde men when they came to the United States, would become known as ''Mr. Missile.'' When Toftoy prepared to leave Redstone Arsenal for a new assignment as commander of Aberdeen Proving Ground in August 1958, the citizens of Huntsville unveiled a monument in his honor at the city's Big Spring Park. Ludy Toftoy retired as a major general and became a missile-industry consultant based in Florida.

One of the great mysteries to emerge from the postwar chaos of Germany was the fate of SS General Hans Kammler, who had risen from obscurity to administrator of the entire V–2 research, development, and production program near the end of the conflict in Europe. Kammler simply vanished—and U.S. counterintelligence operatives apparently made no effort to locate him or determine his fate. At least, on the record they did not.

Reports have lingered for more than 45 years that the ambitious and clever Kammler may have been secretly taken in tow by American operatives, given a new name and identity, and put to work for Uncle Sam in a clandestine role in the Cold War that was raging between the West and the Soviet Union and its Communist allies.

The last known communication from Kammler was sent from his Munich headquarters to his boss, Reichsfuehrer SS Heinrich Himmler, in Berlin, on April 17, 1945. Marked top-secret, the three-sentence message reported that a truck could not be released from Kammler's motor pool. Why would a German general send such an insignificant message to the Reich's second most powerful man? Had Hans Kammler already gone over to the Americans, three weeks before the war ended? Or to the Russians? Was this routine message sent to convey that Kammler was on the job in Munich when, in reality, he had joined with an enemy to save his own skin?

On May 21, 1945, less than two weeks after the German surrender, the Reich's war production boss, Albert Speer, a blood enemy of Hans Kammler (as was Wernher von Braun), was asked by American interrogators for technical details concerning the V–2 rocket. "Ask Kammler," Speer snapped. "He has all the facts and figures."

Curiously, interrogation transcripts have disclosed, the Americans did not even ask, "Who is Kammler?" nor did they pursue the topic. It would appear that the interrogators knew full well who Hans Kammler was, that he had been in full charge of the entire V–2 program since July 1944. Routinely, the Americans could have put in an urgent tracer on Kammler to prisoner of war information units (British and French, as well as U.S.) but presumably no such effort was made.

Long after World War II, Tom Agoston, a British reporter who served for more than a decade as bureau chief in Bonn, Germany, for the International News Service, requested from the British Public Record Office the release of the transcript in which Albert Speer remarked about Hans Kammler as being the man for the Americans to see. Agoston was turned down on the grounds that the documents will remain secret, possibly until the year 2020.

The exchange between Speer and the American interrogators with regard to Kammler is in microfilm copy at the National Archives in Washington, but there is nothing concerning a possible American follow-up to locate the elusive General Hans Kammler, whose whereabouts would seem to have been of paramount importance.

In the months after Germany's surrender, four versions surfaced over Kammler's fate: two of them had him committing suicide, one told of his being killed fighting the Soviet army in Prague, Czechoslovakia, and the fourth had him shot to death by one of his own officers. A junior German officer swore that he had personally buried General Kammler but, strangely, could not recall where.

At any rate, no trace has been found of Hans Kammler—at least as far as the world is concerned—since he sent his message to Reichsfuehrer Himmler on April 17, 1945. Most certainly, Kammler's undeniably brilliant mind had been crammed with technology and German secrets that would have been an invaluable asset to some nation during the long Cold War that threatened to erupt into a shooting war.

Shortly after *Columbia*, carrying history's first moon men, splashed down in the Pacific Ocean in July 1969, NASA's James Webb, who had held the demanding post of administrator for more than eight years, resigned to resume private law practice in Washington, D.C. When Webb had been charged by President Kennedy with flying a man to the moon, how to accomplish the stupendous feat was not even known, nor were the devices to be used. Yet Webb successfully melded NASA, the military, industry, and universities to get the job done—and in an incredibly short timespan.

In February 1970, Wernher von Braun was transferred to Washington, D.C., to be NASA's deputy associate administrator for future planning. Eberhard Rees, von Braun's long-time number-two man, was named to succeed him as director of the Marshall Space Flight Center.

Two years later, von Braun resigned from NASA to accept a position with Fairchild Industries, a Maryland-based aerospace firm, as vice president for engineering and development. He left U.S. government service with a scintillating collection of honors and awards, including: The Department of Defense Distinguished Civilian Service Award, the Department of Army Decoration for Exceptional Civilian Service, the U.S. Chamber of Commerce Award for Great Living Americans, the Distinguished Federal Civilian Service Award (presented by President Eisenhower), the Notre Dame Patriotism Award, the Daughters of the American Revolution Americanism Medal, the NASA Medal of Outstanding Leadership, the American Society of Mechanical Engineers ''Man of the Year'' Award, the Smithsonian Institute's Langley Medal, Associated Press ''Man of the Year in Science'' Award, Freedoms Foundation National Recognition Award, and 26 honorary doctorate degrees from universities.

After leaving NASA, Wernher von Braun continued to speak earnestly of human colonies on the moon and in space, of shuttle flights that would return intact with their crews to earth, of large telescopes in the sky, of huge orbital stations, and of eventual manned flights to Mars, which, every 15 to 17 years, is 35 million miles from earth.

Wernher von Braun, who had been in failing health for two years, died on June 17, 1977, in a hospital in Alexandria, Virginia. At the White House, President Jimmy Carter eulogized the space pioneer as ''a man of bold vision'' and added: ''To millions of Americans, Wernher von Braun's name is inextricably linked to our exploration of space and to the creative application of technology.''[1]

Bernhard Tessman, the German who hid the precious V–2 documents in the Dörnten mine, was among the 89 of the 127 Peenemünde men who remained at the Marshall Space Flight Center until after the moon landing. But Dieter Huzel, who helped Tessman hide the documents, left Huntsville to work for North American Aviation's Space and Information Systems Division in California.

Numerous other Peenemünde experts entering private industry included Wernher von Braun's youngest brother, Magnus, who joined the Missile Division of the Chrysler Corporation; Theodore Buchold with General Electric; Ernst Steinhoff with Rand Corporation; Krafft Ehricke with the Convair Astronautics Division of General Dynamics; and Martin Schilling, who became a vice president of the Raytheon Company.[2]

General Walter Dornberger, the former commander of the Peenemünde rocket experimentation station, joined Bell Aerospace Corporation as a vice president and chief scientist after coming to the United States in the early 1950s. There he was involved in developing the X–1, the world's first rocket airplane, and the rocket-boosted hypersonic glider Dyna-Soar. In his mid–60s, hale and hearty, he retired and returned to Germany to live.

In 1991, Huntsville, which had a population of 16,000 when the Peenemünde rocket team and their families moved there from Texas 41 years earlier, was a booming metropolis of 169,400 people. During the two decades that Wernher von Braun called Huntsville home, his creativity and influence left a rich, research-oriented legacy for the community.

The privately sponsored U.S. Space and Rocket Center was established in Huntsville in 1970 and serves as the archives for von Braun's documents and those of other prominent professionals in the missile and space field. In 1975, Huntsville honored its favorite son by naming the new arts, entertainment, convention, and sports complex the Von Braun Civic Center.

Huntsville remains the home of several dozen members of the original Peenemünde rocket team that migrated from Fort Bliss in 1950. One of them, Walter Wiesman, the youngest of the group, was the recipient of the 1976 Huntsville Distinguished Service Award, and, in 1980, the National Management Association presented him its Medallion for his efforts in promoting the free enterprise system. Wiesman received the George Washington Honor Medal from the Freedoms Foundation at Valley Forge in 1981 and was one of the 1984 recipients of the Brotherhood Award presented by the National Conference of Christians and Jews.

When Neil Armstrong was on the moon in July 1969, he broadcast back to 600 million earthlings: "The responsibility for this flight lies first with history and with the giants of science who preceded this effort." Armstrong could have added that he and other pioneer American astronauts had performed dazzling deeds requiring enormous courage, skill, dedication, and finesse. In barely a decade, these bold young adventurers had taken what the scientists and engineers had provided for them and transformed the seemingly impossible into routine access to space.

The moon landing was Neil Armstrong's last space flight. Later, he taught at the University of Cincinnati and then settled on a farm not far from that city in his native state of Ohio. For many years, he was active in civic and community affairs.

Alan Shepard, the first American in space, retired in 1971 after walking on the moon in *Apollo 14*. The $125,000 he had received from exclusive-story contracts with *Life* magazine and Field Enterprises was the egg that hatched a fortune in real estate. Late in 1971, President Nixon appointed Shepard a delegate to the United Nations' General Assembly.

John Glenn, the first American to orbit the earth, entered the rough-and-tumble field of politics and, after several setbacks, was elected U.S. senator from Ohio on the Democratic ticket in the first post-Watergate election. In 1992, Ohio voters returned Glenn to the Senate for a fourth term.

Frank Borman, a veteran of two space flights, spent a semester at Harvard Business School, went to work for Eastern Airlines and, success-prone as always, rose to be its president.

Edwin "Buzz" Aldrin, the second human on the moon, left the astronaut program in 1971, returned to the Air Force, and retired a year later.

James McDivitt, who had been sent on the midnight ride to Paris by President Lyndon Johnson, became an executive of Consumers Power Company in his hometown of Jackson, Michigan; his companion in that Paul Revere episode, James Lovell, took a job with the Bay-Houston Towing Company in Texas.

Michael Collins, a member of the first moon-landing team, left NASA after that flight and was appointed by President Nixon to be assistant secretary of state for public affairs, a public relations post. In February 1971, Collins was happy to leave the State Department, which he wryly described as "a plush purgatory." A few days later, he jumped across town and became curator of the Smithsonian Institution's National Air and Space Museum.

With the passing of the years, America's pioneer astronauts faded in the public eye. Occasionally, some of them turned up on television or radio commercials: Walter Schirra plugging the railroads, Aldrin Volkswagens, Armstrong banks, and Lovell insurance. Mike Collins claimed he turned down a $50,000 offer to hawk a popular beer.

In the meantime, the public, who had watched in awe and exhilaration at the televised *Apollo 11* mission, began to ask: "What did the moon landing do for *us*?" They knew that Armstrong, Aldrin, and Collins had brought back a lot of rocks, and they had a fuzzy notion that scientists were studying these lunar gems and somehow advancing humankind's knowledge. But all of that was vague and remote to the public, who were unable to relate it to the taxes they had paid to put Americans on the moon.

What the public did not grasp was that from the exhaustive research that produced the rockets, engines, spacecraft, liquid propellants, communications, and spacesuits emerged thousands of spinoffs that produced an extremely broad range of benefits enhancing the lives of virtually every American.

Fallout from space research has vastly improved medical and health care, including many life-saving devices. Other space devices have helped the retarded to learn to speak and the crippled to walk.

Space technology has improved highway, airplane, rail, and automobile safety. New management techniques created by NASA for directing its massive projects have changed the way industry, scientists, engineers, and managers approach virtually any task. What university scientists, professors, and graduate students learn in the pursuit of their research on behalf of space technology feeds back into the teaching and learning process and has a profound impact on American education.

Based on exhaustive studies by NASA and the Space Task Group, President Nixon, on January 5, 1972, announced a new course for the national space effort:

> I have decided that the United States should proceed at once with the development of an entirely new type of space transportation system designed to help transform the space frontier of the 1960s into familiar territory, easily accessible for human endeavor in the 1980's and 1990's.
>
> This system will center on a space vehicle that can shuttle repeatedly from earth to orbit and back. It will revolutionize transportation into near space. It will take the astronomical costs out of astronautics. In short, it will go a long way toward delivering the rich benefits of practical space utilization and the valuable spinoffs from space efforts into the daily lives of Americans and all people.[3]

On August 9, 1974, two and a half years after President Nixon gave NASA the green light to launch the Space Shuttle era, he resigned from office while facing almost certain impeachment by Congress as the result of a series of scandals that came to be known under the umbrella label Watergate. He returned to his home in San Clemente, California, avoided active participation in politics, and spent much of his time playing golf and writing his memoirs.

In 1980, Nixon moved to New York City, and a year later switched his residence to affluent Saddle River, New Jersey. As the years passed and passions began to cool over Watergate, Richard Nixon took on the trappings of an elder statesman, and his views were sought by television network news and other major media.

Harry S Truman, who had given his approval to Operation Overcast in 1945, left the White House in 1953 and returned to his lifelong home in Independence, Missouri, perhaps the only president to leave Washington in the 20th century without being a millionaire. Later, his friends collected funds to build the Harry S Truman Library in Independence. ''Give-'em-Hell-Harry,'' as the scrappy Truman was known, died on December 26, 1972, and was buried in Independence in the Truman Library courtyard.

After the young and vigorous John Kennedy was shot to death in Dallas by an avowed Communist named Lee Harvey Oswald on November 22, 1963, his body was returned to Washington where his flag-draped coffin was placed in the Capitol Rotunda. Throughout the day and night, hundreds of thousands of people passed the guarded casket of the visionary who, two and a half years earlier, had launched a gargantuan program to put an American on the moon.

In one of his first acts, President Lyndon Johnson named the National Aeronautics and Space Administration facility in Florida the John F. Kennedy Space Center.

Although President Kennedy, in May 1961, had made the bold proposal to Congress, a case could be made that it had been Lyndon Johnson, as vice president and chairman of the Space Council, who had conceived the moon-landing caper and proposed it to Kennedy. At any rate, it was Johnson who ordered the project to barrel ahead full blast after Kennedy's assassination, at a time when reports were circulating that Apollo would be scuttled.

After Richard Nixon was sworn into office in January 1969, Johnson retired to his LBJ ranch with a fortune (estimated at $50 million) accumulated since coming to Washington almost broke as a congressman in 1937. In 1971, the Lyndon B. Johnson Library opened at the University of Texas in Austin.

Johnson was stricken with a heart attack and died on January 22, 1973; he was buried in the family cemetery on his Texas ranch. A short time later, the sprawling Manned Spacecraft Center at Houston was renamed the Lyndon B. Johnson Space Center.

Five-star General George C. Marshall, the Army chief of staff who had authorized Operation Overcast in its early stages before it was stamped with presidential authority by Harry Truman, served as secretary of state after World War II and as president of the American Red Cross. When the Korean War broke out in 1950, Truman appointed Marshall secretary of defense, when an act of Congress set aside the law that the post had to be held by a civilian. General Marshall died on October 16, 1959, and is buried at Arlington National Cemetery.

American manned shuttle flights into space continue through 1991 and no doubt will do so far into the future. However, the ongoing argument about the value of manned flights rages on. NASA leaders long have insisted that human crews are vital because even the most sophisticated robots lack the ability to respond to unexpected situations. If the United States is serious about exploring the solar system, they say, unmanned probes to distant celestial bodies must be followed by missions involving humans.

Critics counter that unmanned, expendable rockets can loft most satellites into orbit at far less cost and with much less risk than the reusable shuttle. The argument heated up in the summer of 1991, when Congress began debating whether to fund a $30 billion orbiting space station. NASA planned to use shuttles to ferry up astronauts to assemble the station, then supply it with unmanned rockets. NASA's space station scheme, if funded and successful, would be almost precisely the idea that Wernher von Braun had articulated when he was a teenager experimenting with rockets in Berlin.

No one knows where the exploration of the universe will lead. But the 20th century may well be remembered as the era in which humankind first burst its terrestrial bonds and left planet earth.

# Notes

## CHAPTER 1

1. Joseph P. Kennedy, Jr., an Army Air Corps pilot, was killed in the summer of 1944 while on a bombing mission over German-held France during World War II. Young Joe had been the politically ambitious member in the Kennedy family, and it was said that his father hoped to one day boost him into the presidency.
2. Author interview with Paul B. "Red" Fay, Jr., San Francisco. As assistant secretary of the Navy during the Kennedy administration, Fay remained one of the president's closest friends.
3. Dwight David Eisenhower may well have been lacking in enormous creativity at this time. Although one of history's most popular presidents, he suffered two heart attacks while in office.
4. *Chicago Tribune*, October 3, 1960.
5. *St. Louis Post-Dispatch*, April 7, 1961.
6. *Washington Post*, April 17, 1961.
7. *The New York Times*, May 26, 1961.
8. Theodore C. Sorensen, *Kennedy* (New York: Harper & Row, 1965), p. 526.
9. Ibid.

## CHAPTER 2

1. Polish soldiers, using World War I equipment and arms, fought with exceptional gallantry. On one occasion, a formation of Polish soldiers on horseback charged a German panzer force and was chopped to pieces.
2. The BBC's German Service beamed German language, truthful news to the Third Reich to counter homefront propaganda from the Nazi Ministry of Information. Although Germans were forbidden to listen to the BBC, the broadcast was thought to have had a huge number of listeners.
3. Admiral Wilhelm Canaris' dealings with the British were uncovered by the Gestapo late in the war. After being savagely beaten, Canaris was hanged by a piano wire in Berlin.

4. Hans Heinrich Kummerow was later arrested by the Gestapo and hanged as a spy.
5. Author correspondence with Reginald V. Jones, Aberdeen, Scotland.
6. *The New York Times*, June 17, 1977.
7. Helen B. Walters, *Wernher von Braun* (New York: Macmillan, 1964), p. 51.
8. Ladislas Farago, *The Game of the Foxes* (New York: McKay, 1971), p. 36.
9. Robert H. Goddard died on August 10, 1945, and did not live to see the day of epochal American achievements in space. Many U.S. rocket experts considered Goddard to be the "father of rocketry."

Thirty years after Goddard launched the world's first rocket powered by a liquid propellant motor (in 1926), the site on the outskirts of Auburn, Massachusetts, lay on a public golf course between two fairways. There was no marker or memorial to Goddard's feat until July 13, 1960, when the American Rocket Society unveiled the Goddard Memorial on the site.

10. Farago, *The Game of the Foxes*, p. 36.

## CHAPTER 3

1. Albert Speer, *Inside the Third Reich* (New York: Macmillan, 1952), p. 197.
2. Theodor Morell was considered to be a "quack" by reputable German doctors of that era. Nonetheless, he remained with Adolf Hitler throughout World War II and allegedly steadily increased the potency of the injections as the war turned against the Third Reich.
3. Walter Dornberger, *V–2* (New York: Viking, 1958), pp. 51–53.
4. Ibid., p. 54.
5. David Irving, *The Mare's Nest* (London: Kimber, 1964), p. 23.
6. Figures on Soviet losses in the opening months of Operation Barbarossa would never be known for certain. However, the Germans and the Soviets would agree that the early losses were gargantuan in size.
7. Lend-lease was the description given to the program conceived by President Franklin D. Roosevelt in which the United States shipped scores of old Navy destroyers, weapons, and equipment to besieged Great Britain at token or no cost.
8. Speer, *Inside the Third Reich*, p. 367.
9. *The New York Times*, June 18, 1977.
10. Speer, *Inside the Third Reich*, p. 368.
11. Anthony C. Brown, *Bodyguard of Lies* (New York: Harper & Row, 1975), p. 421.
12. Ibid., p. 432.
13. Dornberger, *V–2*, p. 131.
14. Speer, *Inside the Third Reich*, p. 368.
15. Ibid.

## CHAPTER 4

1. Max Hastings, *Bomber Command* (London: Michael Joseph, 1979), p. 108.
2. Irving, *The Mare's Nest*, p. 40.
3. Author correspondence with Reginald V. Jones, Aberdeen, Scotland.
4. Brown, *Bodyguard of Lies*, p. 349.

5. George Martelli, *The Man Who Saved London* (New York: Doubleday, 1961), p. 214.

6. Joubert de la Ferte, *Rocket* (London: Hutchinson, 1957), p. 48.

## CHAPTER 5

1. Irving, *The Mare's Nest*, p. 117.

2. Speer, *Inside the Third Reich*, p. 369.

3. It was not until 30 years after the close of World War II that the British government even acknowledged the existence of Ultra. It was thought that the ingenious device might be used in a possible war between the Western Allies and the Soviet Union.

4. The Ultra development team was headed by Alan Turing and Alfred Knox, a pair of mathematical geniuses who were considered to be as eccentric as they were brilliant. Although there had never been a threat of the Germans using poison gas, one of the two scientists was taken into custody by a local constable when he was seen walking along a country lane at midnight wearing a gas mask.

5. During the test firing of the V–2, the missile usually landed a mile or two away from the precise target in the marshland 200 miles away. But Wernher von Braun was nearly killed on one occasion when he thought the firing was over for the day and a rocket landed only 100 yards from him as he was strolling toward his private airplane.

6. Willi Ley, *Rockets, Missiles, and Men in Space* (New York: Viking, 1968), p. 210.

7. *V-Weapons (Crossbow) Campaign*, U.S. Air Force Historical Branch, Maxwell Air Force Base, Montgomery, Alabama.

8. Manuscript, John F. Kennedy Library, Cambridge, Massachusetts.

## CHAPTER 6

1. Erik Bergaust, *Reaching for the Stars* (New York: Doubleday, 1960), p. 91.

2. Ibid.

3. Walters, *Wernher von Braun*, p. 72.

4. After World War II, Field Marshal Wilhelm Keitel was tried for war crimes at Nuremberg and was found guilty and hanged.

5. Dornberger, *V–2*, p. 169.

6. Ibid.

7. Ibid., p. 171.

8. Speer, *Inside the Third Reich*, p. 172.

9. Wincenty Hein, *Dora* (London: Kimber, 1962), p. 63.

## CHAPTER 7

1. Author correspondence with Reginald V. Jones, Aberdeen, Scotland.

2. Irving, *The Mare's Nest*, p. 236.

3. On August 19, 1944, Lieutenant Wazny was holed up in a house in a small French town when he was betrayed by a local woman. German soldiers surrounded his hideout and he was shot to death in a gun battle. Three of Wazny's Polish radio operators were captured by the Germans in separate locales.

4. *Impact*, Office of Air Chief of Staff, Intelligence, Pentagon, September 1944.

5. Bohdan Arct, *Poles against the "V" Weapons* (Warsaw: Interpress, 1972), pp. 112–114.

6. Antoni Kocjan was arrested by the Gestapo and executed in August 1944.

7. Jozef Garlinski, *Hitler's Last Weapons* (New York: Times Books, 1978), p. 156.

8. Ibid., p. 158.

**CHAPTER 8**

1. James McGovern, *Crossbow and Overcast* (New York: William Morrow, 1964), p. 78.

2. David Dallin, *Soviet Espionage* (New Haven: Yale University Press, 1955), p. 268.

3. F. H. Hinsley, *British Intelligence in the Second World War* (London: HMSO, 1981), vol. 3, p. 400.

4. Ibid., p. 403.

5. Actually, the first tactical use of the V–2 was not against London but against recently liberated Paris. On September 6, 1944, two days before the first shot exploded in London, two rounds were fired at the French capital. One fell short; the other hit within Paris.

6. Walters, *Wernher von Braun*, p. 76.

7. Before the war concluded, a greater number of V–2s landed in the crucial Allied supply port of Antwerp than in London.

8. *New York Herald Tribune*, January 9, 1945.

9. *The New York Times*, January 9, 1945.

10. *London Daily Express*, January 10, 1945.

11. The Germans' code name *Wacht am Rhein* was designed to convince Allied intelligence that the large number of troop movements was intended to be defensive, not offensive.

**CHAPTER 9**

1. Bergaust, *Reaching for the Stars*, p. 106.

2. *American Weekly*, July 27, 1958.

3. Ibid.

4. NASA Archives, Washington, D.C.

5. The technique for firing a missile from a submerged submarine was later utilized by the United States in its Polaris program.

6. William Shirer, *The Rise and Fall of the Third Reich* (New York: Simon & Schuster, 1971), p. 1104.

**CHAPTER 10**

1. *Time*, March 22, 1945.

2. Tom Bower, *The Paperclip Conspiracy* (Boston: Little, Brown, 1986), p. 93.

3. As a result of his injury, Wernher von Braun had a slightly crooked arm the remainder of his life.

4. Bergaust, *Reaching for the Stars*, p. 110.

5. Joseph Goebbels, *Diaries* (Hamburg: Hoffman and Kampe, 1982), p. 216.

6. Albert Speer, *Infiltration* (New York: Macmillan, 1982), p. 248.
7. McGovern, *Crossbow and Overcast*, p. 36.
8. Tom Agoston, *Blunder!* (New York: Dodd, Mead, 1984), p. 51.
9. *American Weekly*, July 27, 1958.
10. John C. Goodrum, *Space Pioneer* (New York: Strode Publishers, 1982), p. 67.

## CHAPTER 11

1. Author interview with Lieutenant Colonel Haynes Dugan (Ret.), Shreveport, Louisiana.
2. Author interview with Colonel Andrew Barr (Ret.), Washington, D.C.
3. Author interview with Colonel William A. Castille (Ret.), Austin, Texas.
4. Ibid.
5. Report by Major William Bromley, "Evacuation of V–2 Missiles from Nordhausen, Germany," dated July 7, 1945. U.S. Army Military History Institute, Carlisle Barracks, Pennsylvania.
6. *Documents of German History*, New Brunswick, New Jersey, 1958, p. 572.
7. *Combat History of the 324th Infantry Regiment*, Baton Rouge, Louisiana, privately printed, 1946, p. 116.
8. *Time* magazine, April 8, 1949.
9. Walters, *Wernher von Braun*, p. 91.
10. *Time* magazine, April 8, 1949.
11. Ley, *Rockets, Missiles, and Men in Space*, p. 222.
12. Ibid.
13. Colonel General Alfred Jodl was later tried for war crimes in Nuremberg, convicted, and hanged.
14. Fritz Zwicky, *Report on Certain Phases of War Research in Germany*, dated October 1, 1945. Archives of Air Force Museum, Dayton, Ohio.
15. Frederick Ordway III and Mitchell Sharpe, *The Rocket Team* (New York: Crowell, 1979), p. 274.

## CHAPTER 12

1. McGovern, *Crossbow and Overcast*, p. 154.
2. No doubt numerous officers in the Pentagon were aware of the project to bring the 100 V–2s to the United States. But they remained silent so the Pentagon would be in a position to deny involvement or knowledge of the project in the event it backfired and caused international repercussions.
3. Liberty was the classification given to cargo vessels that American shipbuilders produced in rapid fashion in World War II to help the United States supply its global war.
4. Major James Hamill and Major William Bromley received relatively minor decorations for their spectacular intelligence feat.
5. Background and some details from Major William Bromley's 1945 report, "Evacuation of V–2 Missiles from Nordhausen, Germany."
6. McGovern, *Crossbow and Overcast*, p. 166.
7. "History of Army Air Force's Participation in Project Paperclip," August 6, 1948, Maxwell Air Force Base, Alabama.

8. General Lucius D. Clay, *Decision in Germany* (New York: Random House, 1950), pp. 212–213.

9. In postwar years, General Clayton Bissell played a key role in the newly formed Central Intelligence Agency (CIA).

10. Ordway and Sharpe, *The Rocket Team*, p. 283.

11. National Archives, Modern Military 319 Army Intelligence file, box 991.

12. SS General Hans Kammler's ultimate destiny is clouded in mystery. As far as is known, he has never surfaced. One report had it that he had been killed while fighting the Russians in Prague in the closing days of the war.

13. McGovern, *Crossbow and Overcast*, pp. 168–169.

14. Major Robert Staver report, dated May 23, 1946, Research and Development Services, Chief of Ordnance, the Pentagon, U.S. Army Military History Institute, Carlisle Barracks, Pennsylvania.

## CHAPTER 13

1. Years later the United States developed an aircraft based on the Sänger-Bredt concept and called it a ''skip bomber.''

2. Ley, *Rockets, Missiles, and Men in Space*, p. 444.

3. Brown, *Bodyguard of Lies*, p. 364.

4. *London Daily Express*, January 23, 1949.

5. Ibid.

6. Harry S Truman Library, Independence, Missouri, John Franklin Carter file.

7. Ibid.

8. The War Department was the administrative body for the Army and Army Air Corps. When the Pentagon was reorganized in 1949, what had been the War Department became the Department of the Army and the Department of the Air Force, now a separate entity from the Army.

9. Edward R. Stettinius, Jr., a native of Chicago, led the U.S. delegation to the 1945 San Francisco conference, which organized the United Nations. Stettinius died in 1949 at age 48.

10. Record Group, National Archives, Modern Military Section, Walter Dornberger file.

## CHAPTER 14

1. General Lucius D. Clay gained renown by directing the 1948 Berlin Airlift that flew in food and supplies to civilians after the Soviet army blocked roads leading to Germany's largest city. Clay retired in 1949 at four-star rank at age 51.

2. Record Group, National Archives, Modern Military Section, dated September 19, 1945.

3. Ibid., dated September 8, 1945.

4. ''History of Army Air Force Participation in Operation Paperclip,'' Maxwell Air Force Base, Alabama.

5. Developing rockets was but a small fraction of the U.S. Ordnance Department. During World War II, Ordnance spent $45 million and produced 18 million tons of ammunition, 2.5 million trucks, 600,000 artillery pieces, 88,000 tanks, 50,000 self-

propelled weapons, 60,000 jeeps, and countless billions of small-arms ammunition. It would have taken the U.S. automobile industry, at its peak peacetime rate, 15 years to produce the same amount of goods.

6. Record Group, National Archives, Modern Military Section, dated September 26, 1945.

7. Author interview with Walter Wiesman, Huntsville, Alabama.

8. Ibid.

9. Bergaust, *Reaching for the Stars*, p. 129.

10. After arriving in the United States, Wernher changed his first name to Werner. The change was hardly recognized by the media, and his first name continued to be spelled Wernher.

## CHAPTER 15

1. C. Lasby, *Project Paperclip* (New York: Atheneum, 1971), p. 169.

2. In 1946, President Harry S Truman asked Secretary of Commerce Henry A. Wallace for his resignation because of his outspoken criticism of Truman's get-tough policy toward the Soviet Union.

3. Harry S Truman Library archives, Special File, dated December 4, 1945.

4. Ibid.

5. Author interview with Walter Wiesman, Huntsville, Alabama.

6. NASA archives, Washington D.C., filed under "Instructions for German Scientists," dated January 3, 1946.

7. Ibid.

8. Ibid., filed under "Rules and Regulations for Assigned Personnel," dated January 17, 1946.

9. History's first atom bomb was ignited at White Sands Proving Ground in 1945.

10. Ley, *Rockets, Missiles, and Men in Space*, p. 232.

11. Author interview with Walter Wiesman, Huntsville, Alabama.

12. Harry S Truman Library, Henry A. Wallace file, dated January 19, 1946.

13. Ibid., dated January 26, 1946.

14. Ibid., general file, dated January 22, 1946.

15. Ibid., Vannevar Bush file, dated January 24, 1946.

16. General of the Army Dwight D. Eisenhower was called to Washington and succeeded General of the Army George C. Marshall as Army chief of staff on November 19, 1945.

## CHAPTER 16

1. Irmgard Grötrupp's diary, parts of which were published in her 1956 book *Die Besessenen und die Machtigen* (*The Possessed and Powerful*).

2. Lavrenti Beria was no longer Soviet Secret Police chief. Four months after Joseph Stalin died in 1953, Beria was arrested, charged with treason, and executed.

3. Grötrupp, *The Possessed and the Powerful*, p. 167.

4. *New York Times*, October 28, 1946.

5. Ibid.

6. NASA, Washington, D.C., Paperclip file, dated December 4, 1946.

7. Bergaust, *Reaching for the Stars*, p. 21.

8. Harry S Truman Library, Independence, Missouri, Paperclip file, dated December 4, 1946.

9. Ibid., dated December 30, 1946. Others who signed the Alexander telegram were Gordon Allport, Harvard University; Hans A. Bethe, Cornell University; Charles S. Bolte, chairman, American Veterans Committee; W. Russell Bowie, Union Theological Seminary; Preston Bradley, Chicago; Henry Busch, Western Reserve University; Evans Clark, Twentieth Century Fund; Hubert Delany, Domestic Relations Court; Rabbi B. Benedict Glazer, Detroit; Rabbi Solomon Goldman, Chicago; Reverend John Haynes Holmes, New York; Reverend Roy M. Houghton, New Haven; Edward Kasner, Columbia University; Reverend John LaFarge, S. J., New York; W. Hurewicz, Massachusetts Institute of Technology; Solomon Lefschetz, Princeton University; Jermiah T. Mahoney, New York; Bishop Francis J. McConnell, Yale Divinity School; Leonor Michaelis, Rockefeller Institute; Alonzo F. Myers, New York University; Richard L. Neuberger, Portland, Oregon; Bishop G. Ashton Oldham, Albany; Reverend Edward Parson, San Francisco; Reverend Jason Nobel Pierce, San Francisco; Reverend Guy Emery Shipler, New York; Payson Smith, Maine University; Herbert Bayard Swope, New York; Professor Lewis M. Terman, Stanford University; Reverend J. S. Ladd Thomas, Temple University; Harry Trust, Bangor Theological Seminary; Rufus B. von Kleinsmid, University of Southern California; Reverend John C. Walker, Waterbury, Connecticut; and Henry A. Atkinson, co-chairman, Council against Intolerance in America.

10. Harry S Truman Library, Paperclip file, dated March 24, 1947.

11. The *Cold War* phrase was said to have been coined and made popular by Bernard Baruch, an advisor to Harry Truman and several other presidents.

12. Memo to Major General S. J. Chamberlin, Intelligence Director for the War Department, from JIOA Director Bosquet Wev, Records Group, National Archives, dated July 2, 1947.

13. Memo from Brigadier General N. B. Harbold to commanding general of Air intelligence. National Archives, Modern Military History Section, dated July 9, 1946.

14. Walter Dornberger had come to the United States in 1950 as a ward of the Air Force.

15. Memo to Major General S. J. Chamberlin from JIOA Director Bosquet Wev, dated July 2, 1947. National Archives, Modern Military History Section.

16. Background and some details from Irmgard Grötrupp's book *The Possessed and the Powerful*, based on her diary.

**CHAPTER 17**

1. James Vincent Forrestal resigned as secretary of defense due to declining mental health in mid–1949; he committed suicide a few months later.

2. It was the first change in the structure of the military establishment since 1798.

3. It was initially called the Long-Range Proving Ground.

4. Senator John Jackson Sparkman was the Democratic nominee for vice president of the United States in 1952, as a running mate with Adlai Stevenson. The ticket was swamped by Dwight Eisenhower and Richard Nixon.

5. Brigadier General Thomas Vincent was commander of Redstone Arsenal and was responsible for supplying support to the missile group.

6. Author interview with Walter Wiesman, Huntsville, Alabama.

7. Harry S Truman Library, Paperclip file, dated December 21, 1950.

8. Ibid., dated December 28, 1950.

9. *St. Louis Post-Dispatch*, February 22, 1952.

10. *Collier's*, March 22, 1952.

11. *The New York Times*, June 18, 1977.

12. In 1991, Walter Wiesman was still living in Huntsville and lecturing at the University of Alabama–Huntsville.

13. In 1991, Huntsville had more than ten times the population it had when the Fort Bliss group transferred there in 1950.

14. *Huntsville Times*, April 16, 1955.

15. Ibid.

16. A second Charles E. Wilson became secretary of defense in the Eisenhower administration, creating a considerable amount of confusion in the media and government circles. So the second Wilson was dubbed ''Electric Charlie'' (he had been president of General Electric Company), as opposed to ''Engine Charlie'' Wilson, the former General Motors Corporation chairman.

17. The date of the International Geophysics Year was chosen because, 25 years earlier, there had been a South Polar Year, also with international cooperation. And the South Polar Year had been preceded, 50 years before it, by a Polar Year, in which those countries near the Arctic Ocean cooperated in the exploration of the Arctic region.

## CHAPTER 18

1. *New York World-Telegram and Sun*, November 30, 1955.

2. *Atlanta Constitution*, December 7, 1955.

3. *St. Louis Post-Dispatch*, December 7, 1955.

4. *Newsweek*, January 12, 1956.

5. Senator Stuart Symington sought the Democratic nomination for president in 1960 but lost the bid to John F. Kennedy.

6. Stephen E. Ambrose, *Eisenhower* (New York: Simon & Schuster, 1984), p. 313.

7. Ibid.

8. Ibid., p. 347.

9. *New York World-Telegram and Sun*, November 6, 1956.

10. The four-stage Jupiter C was enclosed in what was called the ''launch tub,'' or the ''basket,'' a cylindrical metal container on top of the missile. Prior to lift-off the ''basket'' was rotated rapidly by an electric motor.

11. Colonel John Nickerson's reprimand, signed by Lieutenant General Thomas F. Hickey, commander of the Third Army, said in part: ''Such an arrogant attitude [by you] showed marked disloyalty to your superiors and violated the special trust and confidence reposed in you when you were commissioned an officer.''

12. Two years after the trial, Colonel John Nickerson's security clearance was restored and he held several significant ordnance commands in the years ahead. Nickerson and his wife, Carol, were killed instantly in a New Mexico automobile accident in March 1964.

13. Ley, *Rockets, Missiles, and Men in Space*, p. 315.

## CHAPTER 19

1. Memo of conference, October 8, 1957, Dwight D. Eisenhower Library, Abilene, Kansas.

2. Ibid.

3. The cause of the Vanguard failure was never determined officially. Some scientists believed that it was caused by a leak between the fuel and the rocket engine.

4. *Huntsville Times*, February 1, 1956.

5. *Miami Herald*, February 1, 1956.

6. Anne C. Whitman, dated January 16, 1958, Eisenhower Library.

7. Ibid.

8. Ibid.

9. *Denver Post*, April 5, 1958.

10. *Pittsburgh Sun-Telegraph*, April 15, 1958.

11. Legislative Leaders Meeting, White House, dated March 18, 1958, Eisenhower Library.

## CHAPTER 20

1. Author interview with Kenneth A. Roe, Greenwich, Connecticut.

2. Ibid.

3. George C. Marshall, who retired from the service as a five-star general, was Army chief of staff during World War II and later became President Truman's secretary of state. Marshall is credited with having conceived the Marshall Plan, the largely American-financed reconstruction of Western Europe after the war, a massive aid program designed to keep that region from falling prey to Communist expansion.

4. When Jules Verne wrote his novel in the 1860s, he had named the Gulf of Mexico coast of Florida as the lift-off point for his "moon train." This was only about 100 miles west of Cape Canaveral.

5. Edgar M. Cortwright, ed., *Apollo* (Washington, D.C.: NASA, 1975), p. 11.

6. Ibid.

7. Ham became a national celebrity of sorts. Later the chimp was placed in a zoo where he led a normal and happy life. Mercury workers had become affectionately attached to Ham and often visited him at the zoo.

## CHAPTER 21

1. Alan Shepard retired from the Navy in the 1970s as a rear admiral.

2. *Newsweek*, May 22, 1961.

3. *Christian Science Monitor*, January 31, 1961.

4. John H. Glenn had much in common with President John Kennedy, and they became good friends after Glenn's orbiting flight. Glenn periodically was the president's guest for fishing and for excursions on the presidential yacht.

5. *New York Herald Tribune*, May 25, 1962.

6. *New York Post*, May 24, 1962.

## CHAPTER 22

1. David Wise, a reporter covering the Kennedy tour, recalled the episode years later in *The New York Times*, February 14, 1988.

2. *Houston Chronicle*, September 12, 1962.

3. Transcript of John F. Kennedy speech at Rice, Kennedy Library.

4. *Houston Chronicle*, September 13, 1962.

5. *St. Louis Post-Dispatch*, September 13, 1962.

6. Ibid.

7. Later, McDonnell Aircraft merged with Douglas Aircraft and became known as McDonnell Douglas.

8. A city park in Oradell, New Jersey, Walter Schirra's hometown, was named in his honor.

9. Apollo 204 brought the total of American astronaut deaths to six. Theodore Freeman died when his jet airplane crashed, and Elliott M. See and Charles A. Bassett crashed in a two-seater jet while trying to land at St. Louis Lambert Airport in rainy weather with poor visibility.

## CHAPTER 23

1. Cortwright, *Apollo*, p. 46.

2. *Super Guppy* and *Pregnant Guppy* were built by John M. Conroy, who started with the fuselages of two surplus Boeing C–97s and ballooned out the upper decks enormously to provide a 25-foot clear diameter.

3. *Newsweek*, February 3, 1968.

4. *Toronto Globe and Mail*, January 10, 1969.

5. *Washington Post*, January 10, 1969.

6. The atheist was unsuccessful in the suit against NASA.

7. Frank Borman was appointed deputy director of flight crew operations under Deke Slayton.

## CHAPTER 24

1. Doris Kearns, *Lyndon Johnson* (New York: Harper & Row, 1978), p. 358.

2. Police estimated that some one million people saw the *Apollo 11* lift-off.

3. The Secret Service always used three or more helicopters in carrying a president to an outdoor public function. Anyone intent on killing him by firing a surface-to-air missile would not know in which chopper the president was riding.

4. Neil Armstrong almost did not make it for the *Apollo 11* mission. A few months earlier, he had a brush with death when he had to parachute from a moon-landing training vehicle on which he was practicing.

5. Cortwright, *Apollo*, p. 207.

6. Ibid., p. 209.

7. After the moon-landing mission, Stephen G. Bales received a unique honor for his quick-witted action. At the same time the three moon astronauts were awarded Freedom Medals by President Richard Nixon, Bales received an identical decoration.

8. *Birmingham Post-Herald*, July 23, 1969.

9. Just as America's proposed journey to the moon had its strident detractors and scoffers, so too did Christopher Columbus' plan to go to the New World. A report written by the Talvera Commission, appointed by Spain's Queen Isabella in 1491 to consider the brash young Italian's proposal, declared, "This mission is impossible, vain, and

worthy of rejection. . . . It rests on weak foundations and appears uncertain to any educated person.''

10. Richard M. Nixon, *Memoirs* (New York: Grossett & Dunlap, 1978), p. 429.

## EPILOGUE

1. *Washington Post*, September 3, 1972.
2. Raytheon gained wide public renown many years later as the builder of the Patriot missile that shot down Iraqi scud missiles during the Persian Gulf War in early 1991.
3. *Philadelphia Inquirer*, January 6, 1972.

# Selected Bibliography


Agoston, Tom. *Blunder!* New York: Dodd, Mead, 1984.
Ambrose, Stephen E. *Eisenhower*. New York: Simon & Schuster, 1984.
Arct, Bohdan. *Poles against the V Weapons*. Warsaw: Interpress, 1972.
Bergaust, Erik. *Reaching for the Stars*. New York: Doubleday, 1960.
Bower, Tom. *The Paperclip Conspiracy*. Boston: Little, Brown, 1986.
Brown, Anthony C. *Bodyguard of Lies*. New York: Harper & Row, 1975.
Churchill, Winston S. *Closing the Ring*. Boston: Houghton Mifflin, 1951.
———. *Triumph and Tragedy*. Boston: Houghton Mifflin, 1953.
Clay, General Lucius D. *Decision in Germany*. New York: Random House, 1950.
Collier, Basil. *The Defence of the United Kingdom*. London: HMSO, 1957.
Cortwright, Edgar M., ed. *Apollo*. Washington, D.C.: NASA, 1975.
Dallin, David. *Soviet Espionage*. New Haven: Yale University Press, 1955.
Dornberger, Walter. *V–2*. New York: Viking, 1958.
Farago, Ladislas. *The Game of the Foxes*. New York: McKay, 1970.
Ferte, Joubert de la. *Rocket*. London: Hutchinson, 1957.
Gadney, Reg. *Kennedy*. New York: Holt, Rinehart and Winston, 1983.
Garlinski, Jozef. *Hitler's Last Weapons*. New York: Times Books, 1978.
Goebbels, Joseph. *Diaries*. Hamburg: Hoffman and Kampe, 1982.
Goodrum, John C. *Space Pioneer*. New York: Strode Publishers, 1982.
Hastings, Max. *Bomber Command*. London: Michael Joseph, 1979.
Hein, Wicenty. *Dora*. London: Kimber, 1962.
Hinsley, F. H. *British Intelligence in the Second World War*. London: HMSO, 1981.
Irving, David. *The Mare's Nest*. London: Kimber, 1964.
Kearns, Doris. *Lyndon Johnson*. New York: Harper & Row, 1978.
Lasby, C. *Operation Paperclip*. New York: Atheneum, 1971.
Ley, Willi. *Rockets, Missiles, and Men in Space*. New York: Viking, 1968.
McGovern, James. *Crossbow and Overcast*. New York: William Morrow, 1964.
Martelli, George. *The Man Who Saved London*. New York: Doubleday, 1961.
Miller, Merle. *Lyndon*. New York: G. P. Putnam, 1980.
Nixon, Richard M. *Memoirs*. New York: Grosset & Dunlap, 1978.

Ordway, Frederick, III, and Mitchell Sharpe. *The Rocket Team*. New York: Crowell, 1979.
Shirer, William. *The Rise and Fall of the Third Reich*. New York: Simon & Schuster, 1971.
Sorensen, Theodore C. *Kennedy*. New York: Harper & Row, 1965.
Speer, Albert. *Inside the Third Reich*. New York: Macmillan, 1952.
———. *Infiltration*. New York: Macmillan, 1982.
Taylor, L. B., Jr. *For All Mankind*. New York: E. P. Dutton, 1974.
Tokaev, S. *Stalin Means War*. London: Weidenfeld and Nicolson, 1951.
Walters, Helen B. *Wernher von Braun: Rocket Engineer*. New York: Macmillan, 1964.
Winterbotham, F. W. *The Ultra Secret*. London: Weidenfeld and Nicolson, 1974.
Woolner, Frank. *Spearhead in the West*. Privately printed, 1946.

# Index

**About the Author**

WILLIAM B. BREUER landed with the first assault waves in Normandy on D-Day, then fought across Europe. Later, he founded a daily newspaper in Rolla, Missouri, and after that, a highly successful public relations firm in St. Louis. He has been writing books full time since 1982, twelve of which are now in paperback, and eight of which have become main selections of the Military Book Club.